CMOS PLLs and VCOs for 4G Wireless

CMOS PLLs AND VCOs FOR 4G WIRELESS

ADEM AKTAS
Analog VLSI Lab
The Ohio State University
Columbus, Ohio, USA

MOHAMMED ISMAIL
CTO and Co-Founder,
Spirea AB, Stockholm
* on leave from the Analog VLSI Lab
The Ohio State University
Columbus, USA

Springer Science+Business Media, B.V.

 Electronic Services <http://www.wkap.nl>

Library of Congress Cataloging-in-Publication

Title: CMOS PLLs and VCOs for 4G Wireless
Author (s): Adem Aktas and Mohammed Ismail

ISBN 978-1-4757-8878-5   ISBN 978-1-4020-8060-9 (eBook)
DOI 10.1007/978-1-4020-8060-9

*This book is dedicated to
Gulin & Fatih,
and
Sameha, Ismail, Sr., Tuula,
Ismail, Jr. and Omar*

Contents

Dedication   v
List of Figures   xi
List of Tables   xvii
Preface   xix
List of Acronyms   xxiii

1. INTRODUCTION   1
    1   4G Wireless Terminals   1
    2   4G PLL/VCO Design Challenges   4
    3   Objectives   6
    4   Organization of This book   7
    5   Summary   8

2. OVERVIEW OF VCO/PLL FOR WIRELESS COMMUNICATION   9
    1   Oscillator Overview   10
    2   PLL Frequency Synthesis   12
    3   Phase Noise Specification   17
    4   Summary   20

3. PLL PHASE NOISE ANALYSIS   21
    1   Phase Noise Definition   21
    2   Oscillator Noise Characteristics   25
    3   PLL Noise Analysis: 27
    4   Summary   40

4. BROADBAND VCOs: SYSTEM DESIGN CONSIDERATIONS   41
    1   Radio Architecture and Frequency Planning Considerations   42
    2   CMOS System Integration   44
    3   Summary   50

5. BROADBAND VCOs: CIRCUIT DESIGN CONSIDERATIONS   51
    1   Broadband VCO with Subbands   51
    2   Switching Techniques for Broadband Operation   52
    3   Broadband VCO Implementation   61
    4   Resonator Tank Design   67
    5   Summary   74

6. BROADBAND VCOs: PRACTICAL DESIGN ISSUES   77
    1   PVT Variation Effects on VCO Tuning   79
    2   Tuning Range Calibration (Trimming)   80
    3   External Trimming   80
    4   Auto Calibration (or Self Trimming)   80
    5   Proposed Auto Calibration Technique   82
    6   VCO Pulling In the Integrated Enviroment   85
    7   Summary   91

7. 4GHz BROADBAND VCO: A CASE STUDY   93
    1   Design Objectives   93
    2   Circuit Topologies   94
    3   VCO Tuning   96
    4   Characterization and Measurement Results   98
    5   Summary   106

8. A PLL FOR GSM/WCDMA   107
    1   Frequency Plan and System architecture   107
    2   Dual band VCO Design for Dual Mode Operation   109
    3   Integer-N Architecture   113
    4   Implementation of the Integer-N Architecture   115
    5   Summary   117

9. PLLs FOR IEEE 802.11 a/b/g WLANs   119
    1   Why Multi-band Tri-mode WLANs?   119
    2   Frequency Plan, Radio Architecture and PLL Specifications 120

3	Phase Noise Optimization and Trade-offs	123
4	Loop Filter Design	126
5	The RF (LO1) Synthesizer Implementation	130
6	The IF (LO2) Synthesizer Implementation	139
7	Measured Results	145
8	Summary	145

10. RF CMOS COMPONENT CHARACTERIZATION 149

1	Calibration and Measurement Techniques	150
2	Pad De-embedding	151
3	Pad De-embedding Considerations for RF CMOS	152
4	Probe Pad Layout Techniques	152
5	Measurement Environment Setups	154
6	RF CMOS Test Chip	158
7	Measurement Results and Technology Evaluation	159
8	Summary	163

11. CONCLUSIONS 165

Appendices	167
A S-parameters to Y-parameters Transformation	167
References	169
Index	175

List of Figures

1.1 4G wireless terminals. 2

1.2 4G wireless: convergence and mobility/bit rate trade-off. 3

1.3 Proliferation of short distance wireless within 4G. 4

1.4 A multi-standard device architecture is depicted in a multi-standard wireless network. A single chip SoC is assumed for the baseband/MAC parts. Different radios are used for different standards. 6

1.5 Reverse interconnect scaling in deep sub-micron CMOS [12]. 7

2.1 A typical heterodyne receiver architecture. 10

2.2 Definition of VCO gain or sensitivity. 11

2.3 A typical phase-locked loop architecture. 13

2.4 PLL as a negative feedback model. 13

2.5 (a) PLL linear model (b) Three element passive loop filter. 14

2.6 PLL Bode plots. 16

2.7 LO phase noise specification 19

3.1 The oscillator output spectrum of an ideal oscillator (a) and a practical oscillator (b). 22

3.2 Phase noise definition. 23

3.3 Sidebands around carrier due to (a) AM (b) PM. 25

3.4 VCO phase noise characteristics (a) low-Q case (b) high-Q. 26

3.5 Noise characteristics of a MOS transistor at a fixed bias condition. 26

3.6 PLL noise model. 28

3.7   (a) transfer function for reference. divider, PFD, and CP noises (b) transfer function for the VCO and the control line noises.   30

3.8   PLL bandwidth selection (a) optimum (b) too large (c) too narrow.   30

3.9   Typical PLL output phase noise.   31

3.10   Developed linear PLL model for noise calculation   32

3.11   (a) Divider noise simulation setup (b) Simulated SSB noise of divide-by-8 prescaler circuit.   34

3.12   PFD with charge pump.   34

3.13   The intended open-loop behavior of the PLL.   37

3.14   The actual open-loop behavior of the PLL due to high VCO gain.   37

3.15   Comparison of simulated and measured output phase noise of 4GHz PLL at 3.84GHz for $I_{CP} = 30\mu A$.   38

3.16   The simulated total phase noise of 4GHz PLL at 3.84GHz along with noise contributions from the PLL blocks.   38

3.17   Comparison of simulated and measured output phase noise of 4GHz PLL at 3.84GHz for $I_{CP} = 90u$.   39

3.18   The open-loop behavior of the PLL for $I_{CP} = 90\mu A$.   39

4.1   Receiver architecture for multi-band and multi-standard cellular applications.   42

4.2   Receiver architectures for WLAN applications at the 2.4GHz ISM band.   43

4.3   An integer-N PLL frequency synthesizer architecture.   43

4.4   (a) VCO and CP connections through loop filter (b) Desired operation region for VCO tuning curve (c) CP output current as a function of output voltage.   48

4.5   (a) single continiuos tuning curve (b) tuning curve divided into sub-bands.   49

5.1   Different approaches for broadband VCO with subbands (a) switching inside the LC-tank (b) switching between LC-tanks (c) switching between VCOs.   53

5.2   NMOS transistor as an RF switch (a) VSW is low, the OFF state (b)VSW is high,the ON state.   55

5.3   Differential capacitance switching.   55

5.4   Differential capacitance switching (a) with resistor biasing (b) with MOS biasing.   56

5.5   Band switching circuit (a) simplified schematic (b) Tuning curves.   58

5.6   Band switching circuit (a) simplified schematic (b) Tuning curves.   60

5.7   Inductor switching (a) Single ended inductor switching reported in [51] (b) Differential inductor switching reported in [52].   61

5.8   VCO topologies (a) NMOS only (b) PMOS only (c) NMOS and PMOS complementary.   63

5.9   (a) amplitude correction circuit with detector (b) amplitude correction based on tuning curve selection (c) programmable bias circuit.   66

5.10   Bias filtering (a) conventional low-pass bias filter (b) proposed bias filter.   67

5.11   Bias filtering speed-up (a) power-up delay circuit (b) dynamic delay circuit.   67

5.12   Integrated spiral inductor geometries (a) square (b) octagonal (c) circular   68

5.13   Narrow band inductor SPICE model in ASITIC.   69

5.14   Integrated differential spiral inductor geometries (a) square (b) octagonal (c) circular.   70

5.15   (a) two single-ended inductors used for differential operation (b) a single differential center-tapped inductor.   71

5.16   Varactor structures in CMOS technology and C-V characteristics (a) p+ to n-well junction varactor (b) NMOS varactor (c) Accumulation mode MOS varactor (d) PMOS varactor.   73

5.17   (a) test bench for simulation (b) simulated C-V curves for different AC signal amplitudes.   75

6.1   (a) VCO with control inputs. (b) VCO tuning curves divided into subbands. (c) CP output current as a function of output voltage.   78

6.2   Conventional auto calibration architecture.   81

6.3   The calibration frequency and control voltage.   82

6.4   Proposed auto calibration architecture.   83

6.5   Flow chart of operation in the proposed calibration architecture.   84

6.6   The reference voltage circuit.   84

6.7   A typical simplified TDD radio architecture.   86

6.8   VCO with its inputs and outputs   87

6.9   (a) RF Transmitter (b) ramped power-up timing scheme.   90

7.1   Simplified complementary NMOS and PMOS VCO circuit schematic.   94

7.2   Simplified PMOS VCO circuit schematic.   95

7.3   PMOS capacitance switching circuit (a) simplified schematic (b) layout.   97

7.4   MIM capacitance switching circuit.   97

7.5   Accumulation mode MOS varactor (a) physical cross-sectional layout (b) C-V characteristic (c) schematic symbol.   98

7.6   Simulated characteristics of accumulation mode varactor (a) C-V characteristics (b) Series resistance versus tune voltage (c) Q versus tune voltage.   99

7.7   (a) test structure of the differential inductor (b)comparison of measured and simulated Q of inductor.   100

7.8   Current density on the differential inductor at 4GHz.   100

7.9   Measured tuning curve of VCO1 for different trim inputs.   101

7.10   Measured tuning curve of VCO2 for different trim inputs.   102

7.11   Comparison of tuning curves for VCO1 and VCO2.   102

7.12   Measured and simulated phase noises (a) VCO1 (b) VCO2.   104

7.13   Measured supply sensitivities of (a) VCO1 (b) VCO2.   105

7.14   Die photo of the 4GHz VCO in the PLL test prototype.   106

8.1   Proposed GSM/WCDMA frequency synthesizer plan for wide band IF double conversion operation.   108

8.2   Frequency synthesizer architecture.   109

8.3   Simulation result of the VCO control voltage transient response.   110

8.4   The simplified dual-band VCO schematic view.   111

8.5   The simulated phase noise of the VCO.   112

8.6   Tuning versus control voltage simulated results for each mode. 113

8.7   System block diagram of the Integer-N divider.   114

8.8   Layout view of the whole synthesizer.   117

9.1   Wireless LAN growth [75].   120

9.2   Frequency plan for multi-band and multi-mode operation WLAN applications.   122

9.3   Wideband-IF radio transceiver architecture.   122

9.4   Quadrature LO generation in a zero-IF (or low-IF) receiver using LO1 and LO2. LO1 is fixed while LO2 is used for frequency synthesis (or vice versa).   124

9.5   A typical PLL SSB phase noise profile.   125

9.6   Phase noise requirements for VCO and PLL close-in noises for a constant integrated noise value of $L_{tot} = -33dBc$ for different values of B.   127

9.7   Loop filter architecture (a) conventional loop filter architecture (b) proposed loop filter architecture.   129

9.8   The RF (LO1) Frequency synthesizer architecture.   131

9.9   Simplified schematic of 4GHz VCO.   132

9.10   Equivalent circuit model used to represent mixers' LO input.   134

9.11   Simplified VCO buffer schematic.   135

9.12   VCO buffer trimming capacitance schematic.   136

9.13   (a) differential line structure (b) simulation model.   136

9.14   RF prescaler architecture.   137

9.15   (a) divide-by-2 schematic (b) CML latch.   138

9.16   (a) divide-by-4 schematic (b) medium speed latch.   139

9.17   The IF (LO2) Frequency synthesizer architecture.   140

9.18   2-stage PPF schematic.   142

9.19   IQ phase error due to 2nd and 3rd harmonic levels at PPF output.   143

9.20   Dual-modulus prescaler architecture.   144

9.21   Composite phase noise plot of the PLLs at the 2.4GHz transmitter output.   145

9.22   Die photo of (a) LO1 (b) LO2.   146

9.23   2.4GHz transmitter performance with 64-QAM signal.   147

9.24   5GHz transmitter performance with 64-QAM signal.   147

10.1   2-port network analyzer measurement.   150

10.2   Device and dummy layout.   151

10.3   a) A conventional G-S-G probe pad layout (top view) and a simple equivalent model b) Ground shielded G-S-G probe pad layout and a simple equivalent model.   154

10.4   A cross sectional view of the lossy 2-port structure and coupling path between ports through substrate.   154

10.5   a) One port G-S-G top view of the conventional probe pad layout (for 3 metal CMOS process, M1,M2, and M3 are metal layers). b) side view of the layout. c) A simple equivalent circuit for S to G path. d) s_{11} parameter plot on the Z-smith chart.   155

10.6   a) One port G-S-G top view of the ground shielded probe
          pad layout (for 3 metal CMOS process, M1,M2, and M3
          are metal layers). b) side view of the layout. c) A simple
          equivalent circuit for S to G path. d) s_{11} parameter plot
          on the Z-smith chart.   156
10.7   2-port and 1-port inductor measurement setup on die.   157
10.8   Differential varactor measurement setup.   158
10.9   RF MOS device measurement setup.   159
10.10   The layout floor plan for the test chip.   160
10.11 S_{11} of the open structure.   161
10.12 Q of the open structure.   161
10.13 S_{11} of the test inductor structure, IND2.   162
10.14 Q of the test inductor structure, IND2.   162
10.15 Q of the test inductor structure, IND3.   163

List of Tables

1.1	Table of short range wireless standards.	5
7.1	Comparison of VCO1 and VCO2 performance parameters.	103
8.1	Specifications of interest in GSM and CWDMA standards for synthesizer frequency planning.	109
8.2	Summary of the dual-mode synthesizer configuration.	110
8.3	Required tuning range specifications.	112
8.4	Summary of VCO performance.	112
8.5	State transition table for 4-bit counter.	116
9.1	Summary of the RF (LO1) and IF (LO2) PLL specifications.	123
9.2	PLL close-in noise, VCO noise at 1MHz offset, and PLL noise floor for different values of loop bandwidth.	126
9.3	Simulated characteristics of LO1 VCO.	133
9.4	Simulated characteristics of the IF (LO2) VCO.	141
9.5	Comparison of this work with published works.	147

Preface

As we move to the next millennium, the telecommunication needs of tomorrow are not very clear. Present day telecommunication services are dominated by voice. Broadband access to the Internet and web browsing are growing rapidly. If the past decade is any indication, wireless communication will continue to grow as the demand for mobility continues to surpass all expectations. Most forecasts predict that the number of mobile phones worldwide will exceed one billion by 2005!

At the time of writing this book, mobile operators in many countries start to provide third generation (3G) Wireless Wide Area Network (WWAN) services (e.g Swedish operators http://www.tre.se, telia.se and tele2.se) adding video and multimedia applications. Beyond 3G comes 4G where there is no agreed upon definition for 4G mobile systems. However beyond 3G, wireless systems will consist of a combination of different access technologies namely cellular systems (existing 2G and 3G for WWAN), Wireless Local Area Networks (WLANs,802.11 a/b/g) and Wireless Personal Area Networks (WPANs, Bluetooth).

These access systems will be connected via a common IP-based core network that will also handle "network convergence" providing seamless handover among these different networks. As a result mobile terminals must be able to work across different standards supporting both convergence as well as migration from existing to emerging higher data rate systems, whether cellular or WLAN.

Currently and with the emergence of WLAN for higher data rates at shorter distances for "hot-spots", a debate is ensued as to whether cellular and WLAN technologies are competing or complementary. WLANs are growing very rapidly and provide the potential for carrying voice over the Internet (VoIP) with voice over WLAN (VoWLAN) technology at much cheaper rates as, unlike cellular, they are based on the unlicensed ISM and UNII bands.

Central to this debate however is the success, or lack thereof, of developing low power cost effective multi-standard multi-band chip-set solutions, particularly the radio part of the solution, catering for either or both technologies. This is particularly true for WLAN as it moves from being PC-centric into being handheld-centric. Higher levels of integration in deep sub-micron CMOS technologies promise to propel us beyond 3G with many new wireless products and innovative applications. Over the past few years, researchers in the field of wireless semiconductor began to report chip design solutions for multi band multi standard wireless applications. Also, engineers at industry have succeeded with development of CMOS single chip multi-standard radio transceiver and digital baseband chips.

This book is devoted to the subject of CMOS phase lock loops (PLLs) and voltage controlled oscillators (VCOs) design for future broadband 4G wireless devices. These devices will be handheld-centric, requiring very low power consumption and small footprint. They will be able to work across multiple bands and multiple standards covering WWAN (GSM,WCDMA), WLAN (802.11 a/b/g) and WPAN(Bluetooth) with different modulations, channel bandwidths, phase noise requirements, etc. As such, the book will discuss design, modeling and optimization techniques for low power fully integrated broadband PLLs and VCOs in deep sub-micron CMOS. To our knowledge, this is the first book on the subject.

First, the PLL and VCO performances are studied in the context of the chosen multi-band multi-standard radio architecture and the adopted frequency plan. Next a thorough study of the design requirements for broadband PLL/VCO design is conducted together with modeling techniques for noise sources in a PLL and VCO focusing on optimization of integrated phase noise for multi-carrier OFDM 64-QAM type applications. Design examples for multi standard 802.11a/b/g as well as for GSM/WCDMA are fully described and experimental results from 0.18μm CMOS test chips have demonstrated the validity of the proposed design and optimization techniques. Equally important the work describes techniques for robust high volume production of RF radios in general and for integrated PLL/VCO design in particular including issues such as supply sensitivity, ground bounce and calibration mechanisms.

This book is intended for use by graduate students in electrical and computer engineering as well as RFIC design engineers in both the wireless and the semiconductor industries. It will also be useful for design managers, project leaders and individuals in the wireless semiconductor marketing and business development.

Chapter 1 provides introduction material and discusses the objectives as well as the organization of different chapters.

Chapter 2 provides an overview of VCO and PLL design for fully integrated single chip radio solutions.

Chapters 3 addresses phase noise in broadband PLLs.

Chapters 4, 5 and 6 discuss different aspects of broadband VCO design starting with system level considerations in Chapter 4, circuit design issues in Chapter 5 and practical design considerations in Chapter 6.

Chapter 7 presents a case study of a CMOS 4GHz VCO and discusses its applications.

Chapter 8 deals with a case study for PLL/VCO design for cellular multi-standard applications while Chapter 9 presents a case study for designing PLLs and VCOs for a multi band multi standard CMOS radio transceiver fully compliant to the 802.11a/b/g WLANs.

Chapter 10 discusses RF CMOS characterization and measurements techniques used throughout this work including pad-dembedding and device characterization. Chapter 11 provides a summary of the contributions of this work.

This book has its roots in the Ph.D thesis of the first author. We would like to thank all those who assisted us at different phases of this work including our colleagues at the Analog VLSI Lab, The Ohio State University and at Spirea AB, Stockholm and Spirea Microelectronics, Dublin, Ohio. We would like to specially thank Yiwu Tang, Fredrik Johnsson, Rami Ahola, Laurent Nouger, Martin Sanden, Peter Olofsson, James Wilson, Kishore Rama Rao. We certainly like to thank our families for their understanding and encouragement during the development of this work. We would not have completed this book without their support.

ADEM AKTAS AND MOHAMMED ISMAIL

Columbus, Ohio

List of Acronyms

2G	2nd generation
3G	3rd generation
4G	4rd generation
ADC	Analog-to-Digital Converter
AFC	Automatic Frequency Control
AGC	Automatic Gain Control
AM	Amplitude Modulation
BB	baseband
BER	Bit Error Rate
BFSK	Binary Frequency Shift keying
BPSK	Binary Phase- Shift Keying
CDMA	Code Division Multiple Access
CML	Current Mode Logic
CP	Charge Pump
dBc	dB with respect to the carrier
DCR	Direct conversion Receiver
DFF	D-type Flip-flop
DMP	Dual Modulus Prescaler
DSB	Double Sideband
DSP	Digital signal processing
EM	Electromagnetic
EVM	Error vector Magnitude
FDD	Frequency-division duplexing
FM	Frequency Modulation
GFSK	Gaussian Frequency Shift Keying
GMSK	Gaussian Minimum Shift Keying
GPS	Global Positioning System
GSM	Global System for Mobile communications
IF	Intermediate Frequency

IP	Internet Protocol
ISI	Intersymbol Interference
LNA	Low-Noise Amplifier
LO	Local Oscillator
LPF	Low-Pass Filter
MAC	Medium Access Control
MIM	Metal-insulator-metal
OFDM	Orthogonal Frequency Division Multiplexing
PFD	Phase-Frequency Detector
PLL	Phase-locked Loop
PM	Phase Modulation
PPF	Poly-Phase Filter
PVT	Process, supply voltage, and temperature
QAM	Quadrature Amplitude Modulation
QPSK	Quadrature Phase Shift Keying
RF	Radio Frequency
rms	Root-Mean Square
RX	Receiver
SAW	surface acoustic wave
SNR	Signal-to-Noise Ratio
SSB	Single Sideband
TDD	Time-division duplexing
TX	Transmitter
VCO	Voltage-Controlled Oscillator
VoIP	Voice over IP
VoWLAN	Voice over WLAN
WAN	Wireless Area Network
WCDMA	Wideband CDMA
WLAN	Wireless Local Area Network
WPAN	Wireless Personal Area Network

Chapter 1

INTRODUCTION

The mobile wireless communication industry has seen explosive growth in the last decade of the twentieth century. Migration to high frequencies and wider bands for radio access allowed increased capacity for more users. This coupled with increased demand for high bandwidth wireless communication has resulted in rapid commercial development of wireless devices. This rapid development has generated cost effective solutions and wide spread usage among the public (405 million cell phones were sold in 2002) [1]. Higher levels of integration with technological advancements in CMOS have allowed the implementation of complex digital modulation schemes with reasonable power consumption, cost, and size for higher data rates. This, in turn, has paved roads for today's modern mobile wireless systems (second and third generations: 2G and 3G) with high data rates at reasonable cost. This chapter outlines a scenario for fourth generation (4G) wireless systems and discusses the challenges for designing phase-lock loops (PLLs) and voltage-controlled oscillators (VCOs) needed for 4G. Then, the chapter discusses the objectives and outlines the organization of the book.

1. 4G Wireless Terminals

There is no unanimously accepted definition for 4G wireless systems [2, 3]. 4G systems will likely consist of a layered combination of different access technologies:

- Cellular systems (e.g. existing 2G and 3G systems for wide area network or WAN)

- Wireless LANs (e.g. IEEE 802.11 a/b/g for dedicated indoor applications at hotspots)

- Personal LANs for short range personal area networks (PAN) and low mobility applications (e.g. Bluetooth, IrDA, etc.) around a room in the office.

To support such a scenario, Figure 1.1 depicts the functionalities or chipsets needed in a 4G wireless terminal. Global positioning system (GPS) is also needed for location-aware applications such as in emergencies.

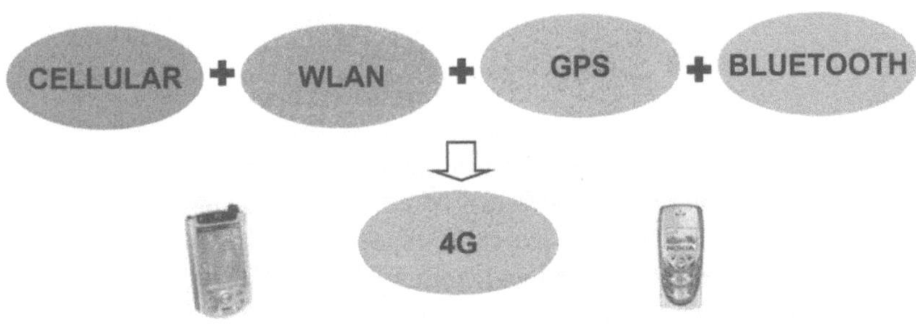

Figure 1.1. 4G wireless terminals.

These access systems will be connected via a common IP-based core network that will also handle working between the different systems [3]. The core network will enable inter and intra access handover. The European countries are also considering digital video and audio broadcasting via the common IP network.

The bit rates of 3G systems are around 10 times more than 2G systems. Figure 1.2 shows the mobility and bit rate trade-off. It also illustrates the evolution of wireless systems achieving migration to higher data rates with backward compatibility to existing wireless infrastructure as well as convergence of wireless standards in 4G with seamless handover for "always best connected" service. 4G systems may provide 10 times higher data speeds relative to 3G systems [3]. They will support bit rates of 2Mbps for vehicular and over 50Mbps for in-door

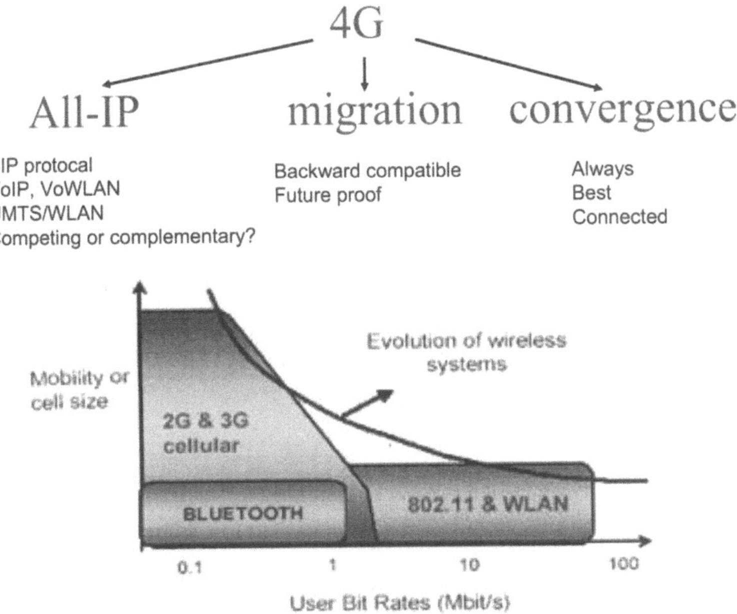

Figure 1.2. 4G wireless: convergence and mobility/bit rate trade-off.

applications. 4G systems are expected to meet the requirements of next generation Internet protocols.

4G wireless systems require new approaches for systems design. The following key aspects of mobile systems design present challenges [3]:

- Multiple access scheme (CDMA may require broadband spreading, 100MHz; TDMA will need high-speed equalizers; OFDM will need linear amplifiers.)

- Multiple antennas (smart antennas)

- IP header

- MAC layer

- Multiple bands and systems (The ability to access various types of access systems that use different RF and basebands will need software radios for both RF and baseband).

While 3G cellular systems are currently being deployed, WLAN technology based on unregulated bands (e.g. the 2.4GHz ISM and the 5GHz UNII bands) is witnessing tremendous growth. Combined with voice over IP (VoIP) technology, WLAN has the potential for becoming the technology of choice for low

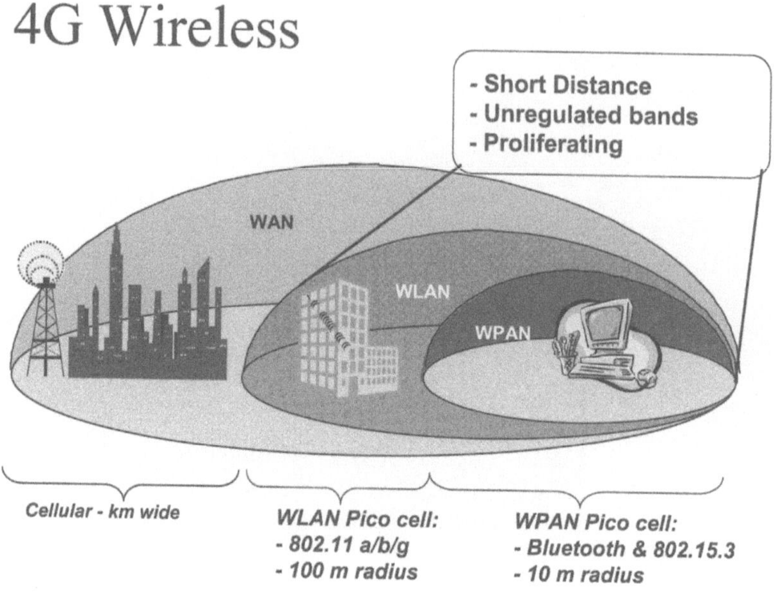

Figure 1.3.   Proliferation of short distance wireless within 4G.

cost mobile internet telephony. This will be helped by introducing quality of service (802.11e) to WLAN currently being ratified by the IEEE. Consequently, we see major proliferation of short distance wireless technology in general and of WLAN in particular (Figure 1.3). Table 1.1 lists a summary of emerging short range wireless standards.

2.   4G PLL/VCO Design Challenges

The wide spread use of mobile wireless terminals in the last decade has lead to demand for radio transceivers with low cost, small-form factor, and low power. Furthermore, mobile terminals are required to operate in multiple RF standards to serve new generations of standards while being backward compatible with existing standards. This has led to a demand for multi-standard/band radio operation which requires the ability to process signals corresponding to wide range of frequency bands and channel bandwidths with different modulation schemes. In addition to these requirements, low phase noise and high performance are required by digital modulation techniques which are used in new generations of wireless standards for efficient use of the available frequency spectrum. While digital basebands and MACs can benefit from digital system-

Standards		Frequency Range	Multiple Access	Modulation	Data Rate
IEEE 802.11 WLAN Standards	2.4GHz DSSS	2.4GHz	DSSS	DBPSK DQPSK	1Mbps 2Mbps
	2.4GHz FHSS	2.4GHz	FHSS	2GFSK 4GFSK	1Mbps 2Mbps
	2.4GHz DSSS High Data Rate	2.4GHz	DSSS	CCK	5.5Mbps 11Mbps
	5/2.4GHz OFDM	5.25-5.8GHz /2.4GHz	OFDM	BPSK/QPSK 16/64 QAM	6,9,12,18 24,36,48,54
HIPERLAN/2		5.250GHz 5.600GHz	OFDM	BPSK/QPSK 16/64 QAM	6,9,12,18 27,36,54Mbps
Bluetooth™		2.4GHz	FHSS	2GFSK	1Mbps
UWB		3-10GHz	OFDM	QAM	500Mbps

Table 1.1. Table of short range wireless standards.

on-chip technology, radio chips currently deployed in wireless terminals are still fragmented (see Figure 1.4), i.e. different radios for different standards. A common single chip multi-band/multi-standard radio solution should result in the best performance/price ratio. All these, in turn, require a high performance broadband frequency synthesizer solution. A straightforward approach for the implementation of frequency synthesis is to use a distinct or separate synthesizer for each standard or frequency band. This approach is often called stacked-radio approach (Figure 1.4) and is not well suited for high levels of integration, with optimized performance/price ratio and with low levels of power consumption.

The large scale integration and RF capabilities of deep sub-micron CMOS technology can be used for high levels of mixed signal radio integration for multi standard operation. The design and analysis of RF transmitter and receiver circuits in CMOS technology are discussed in recent literatures [4–6]. In order to optimize the integration of multi-standard/band radio systems, the transceiver architecture and the synthesizer must be designed concurrently to maximize hardware share among operating modes, with particular attention paid to frequency planning [7, 8]. In addition to a careful frequency planning, a

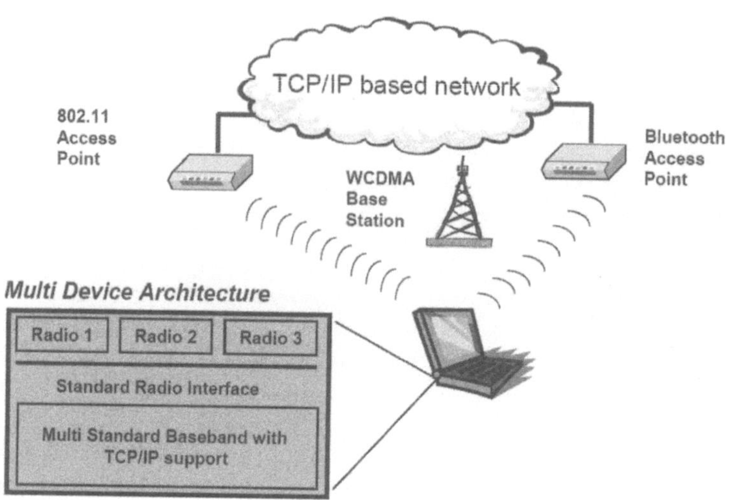

Figure 1.4. A multi-standard device architecture is depicted in a multi-standard wireless network. A single chip SoC is assumed for the baseband/MAC parts. Different radios are used for different standards.

broadband PLL frequency synthesizer design in CMOS necessitates high performance circuit and architecture implementation of the following blocks; 1) low phase noise, low power, broadband RF VCO 2) quadrature LO signal generation 3) high-speed divider (prescaler) 4) RF buffer and limiter. Early research efforts for the monolithic CMOS VCO had focused on the design and implementation of high-Q resonator components, especially integrated inductors, to achieve low phase noise at an acceptable level of power consumption [9–11]. Modern CMOS technologies offer thick metal layers on a high-resistivity substrate to implement high-Q inductors (see Figure 1.5 [12]). The two important challenges to implement RF VCOs in a fully integrated RF transceiver are; i) on-chip noise coupling through cross-talk between blocks and substrate ii) CMOS fabrication tolerances (process variation). Fully integrated PLL frequency synthesizers solution must address top level implementation issues as well as individual block design issues.

3.   Objectives

This book investigates and explores broadband PLL frequency synthesizer implementation in CMOS technology with focus on the integration of broadband VCOs. The specific topics covered in this book include:

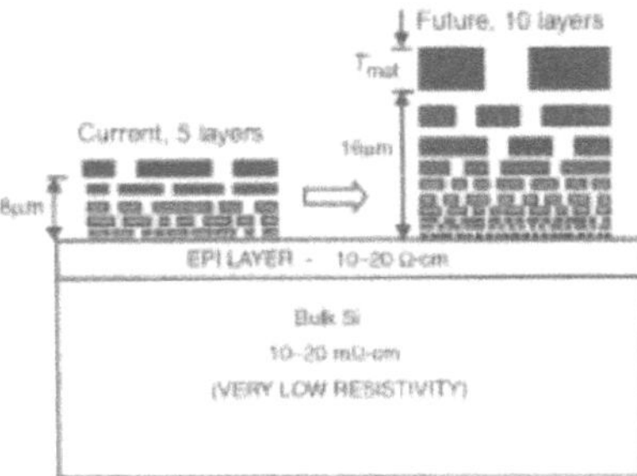

Figure 1.5.   Reverse interconnect scaling in deep sub-micron CMOS [12].

- PLL phase noise analysis and optimization: Stringent requirements on phase noise in most wireless standards represent a great challenge for integration of a frequency synthesizer in CMOS technology.

- Development of innovative PLL frequency synthesizer architectures for multi-band and multi-standard radio solutions.

- Design and development of the monolithic broadband VCOs with low phase noise and power consumption in CMOS technology.

- Investigate the use of the capabilities of low cost digital signal processing (DSP) of CMOS technology to compensate for impairments in analog blocks through auto calibration or trim circuits.

- RF CMOS component characterization using on wafer microwave probing with a main focus on integrated inductors and varactors.

The applications of the broadband PLL frequency synthesizer are also demonstrated by designing and implementing; i) A fully integrated dual-mode frequency synthesizer for GSM and WCDMA standards in 0.5μm CMOS technology. ii) Frequency synthesizers for an IEEE 802.11a/b/g WLAN radio transceiver in 0.18μm CMOS technology.

4.   Organization of this book

The organization of the book is summarized in this section.

Chapter 2 overviews oscillators and PLL frequency synthesizers in mobile communications systems.

Chapter 3 treats the analysis and optimization of PLL phase noise. Simulation results from noise models based on the analysis are demonstrated by characterizing the noise contribution of individual PLL blocks.

Chapter 4 discusses the system design considerations for broadband RF CMOS VCOs in fully integrated transceiver solutions. Multi-band radio receiver architectures based on a single broadband VCO are discussed.

Chapter 5 deals with circuit design considerations for broadband RF CMOS VCOs as well as VCO implementation in fully integrated transceiver solutions. RF VCO with subbands is considered for broadband implementation.

Chapter 6 deals with practical VCO design issues due to process, supply and temperature variations. Also, VCO pulling in an integrated environment is discussed. Circuits and auto calibration techniques are presented for robust VCO design. Techniques and trade-offs, at both the circuit and system levels, for minimizing VCO pulling are also discussed briefly.

Chapter 7 is devoted to the design of a 4GHz broadband VCO in 0.18μm CMOS technology. PMOS and CMOS active circuit topologies were evaluated for performance. The PMOS topology is chosen for better noise performance.

Chapter 8 describes the implementation of a fully integrated dual-mode frequency synthesizer for GSM and WCDMA standards in a 0.5μm CMOS technology. Integer-N and fractional-N PLL architectures are used for the frequency synthesizer to optimize the performance while using a single loop filter.

Chapter 9 describes implementation of two PLL frequency synthesizers, used in a fully integrated multi-band/standard WLAN radio transceiver in a 0.18μm CMOS technology for IEEE 802.11a/b/g standards at the 2.4GHz and 5.25GHz frequency bands. Frequency planning and architecture design are optimized for maximum hardware sharing, low power dissipation, and ease of the implementation in CMOS technology. Phase noise optimization and loop filter design of the synthesizer are presented. Implementations of the RF and the IF PLL synthesizers are described at the circuit and architecture level as well as important CMOS design and integration issues.

In Chapter 10, RF CMOS component characterization methodologies are presented. Measurement and calibration techniques are also investigated in this chapter. Chapter 11 summarizes the results and contributions presented in this book.

Appendix A describes S-parameters to Y-parameters transformation.

5.   Summary

This chapter presents the evolution of wireless systems beyond 3G. A scenario for 4G wireless is discussed where migration to higher data rates and convergence of cellular technology with short range wireless standards, particularly WLAN, are emphasized. Challenges facing chip-set development for 4G are also discussed with particular emphasis on radio and frequency synthesizers. The chapter also provides a discussion on the objectives and the organization of this book.

Chapter 2

OVERVIEW OF VCO/PLL
FOR WIRELESS COMMUNICATION

PLL frequency synthesizers are important building blocks in wireless communication systems. Frequency translation in the RF transceiver circuits requires accurate, programmable, high performance local oscillator (LO) signals. The quality of LO signal plays a key role for the overall performance of the wireless systems by determining how closely channels can be placed in narrow-band systems and how closely constellation points can be placed in I/Q plane in digital modulated systems. The first one is related to the phase noise performance at a given offset frequency while the second one is related to the integrated phase noise performance over the signal bandwidth. The phase noise performance of a PLL synthesizer depends on both individual block performance parameters and PLL system behavior. Therefore, the design and implementation of a high performance PLL synthesizer require expertise in broad range of fields; RF, analog, and digital circuit design as well as system level design.

Figure 2.1 shows a typical heterodyne receiver used in wireless communication systems. The RF LO signal in the receiver side is used to translate the incoming RF signals to lower IF frequency range [5, 11]. The channel selection can be performed either at RF LO or IF LO depending on the arrangement of the architecture. Fully integrated solutions often employ zero-IF radio architecture which converts the incoming RF signal directly to baseband frequency range with a single LO. The zero-IF architecture avoids IF filters and hence allows fully integration of RF front-end functionality [11]. The transmit up-conversion is also performed in a double conversion scheme as shown in Figure 2.1. The RF and IF LO signals can be shared between the receiver and transmitter in time-division duplexing (TDD) systems.

This chapter briefly overviews the oscillators and PLL frequency synthesizers for wireless applications. Phase specification for a LO signal is also discussed.

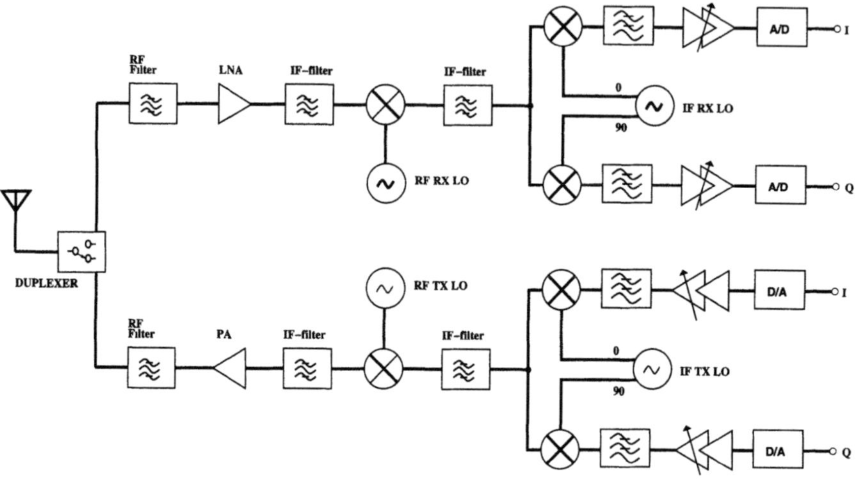

Figure 2.1.   A typical heterodyne receiver architecture.

1.   Oscillator Overview

Oscillators are key components in implementing PLL frequency synthesizers. Each PLL employs two oscillators; a crystal oscillator as reference signal and a RF VCO for synthesized output signal. A typical crystal oscillator exhibits excellent accuracy and very high performance due to high-Q resonator. However, a crystal oscillator cannot be built at the RF frequency range and cannot be synthesized to be used as a LO signal. RF VCO with a resonator LC-tank or transmission lines can be built at the RF frequencies and can be made tunable, but they exhibit very poor performance and they are unstable due to low-Q resonator. A PLL frequency synthesizer employing a crystal oscillator as a reference and a VCO as an output can generate high performance, accurate, and synthesize-able LO signal. Analysis and design of RF and microwave oscillators have been extensively studied, and can be found in references [11, 13, 14].

Voltage-controlled Oscillator (VCO)

Wireless applications require that oscillators be "tunable", i.e, their output frequency be a function of a control input. Usually, a voltage signal is used for control input, and these oscillators are called voltage-controlled oscillator (VCO). Current controlled oscillators (CCO) are also feasible but are not widely used in RF systems because of difficulties in tuning the high-Q resonators by means of current.

An ideal voltage-controlled oscillator has an output frequency which is a linear function of its control voltage. The output frequency of an ideal VCO can be expressed as;

$$\omega_{out} = \omega_o + K_{VCO}V_{cont} \tag{2.1}$$

where ω_o is *the free running frequency* at $V_{cont} = 0$ and K_{VCO} denotes the *gain* or *sensitivity* of the oscillator in MHz/V. The achievable output frequency range, $\omega_2 - \omega_1$, is called *tuning range* as shown in Fig. 2.2.

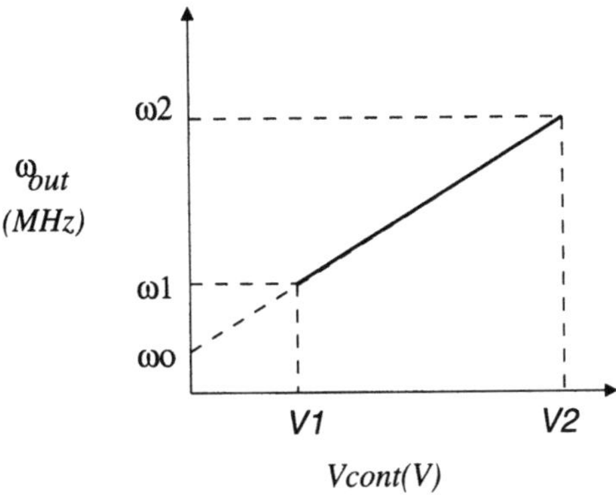

Figure 2.2.   Definition of VCO gain or sensitivity.

Some of the important performance parameters of a VCO are summarized as follow:

Tuning range: The desired tuning range for a fully integrated VCO is dictated by two specifications: (1) the operating frequency range of the application (2) the variation of the VCO output frequency for corresponding control voltages with process and temperature. The second factor imposes great challenges in fully integrated CMOS VCO implementation. CMOS VCOs are often designed with a tuning range two times wider than the desired application tuning range in order to compensate for process and temperature variations.

An important consideration in the design of VCOs is the noise on the control line which can vary the output phase and frequency of the circuit. As the gain of the VCO increases, the sensitivity of the VCO to control line noise increases. To minimize the effect of control line noise, the VCO gain must be as small as possible, which implies that tuning range must be minimized. This conflicts with the second requirement of a wide tuning range.

Tuning linearity: An ideal VCO has an output frequency which varies linearly with control voltage. However, practical VCOs exhibit a nonlinear tuning characteristic, i.e. K_{VCO} is not constant along the tuning range. Linear tuning is desired to keep the VCO gain constant over the tuning range so that the PLL performance does not change while operating at different frequencies.

Output Amplitude: A large output amplitude is desired for lower phase noise. However, larger output amplitude implies increased current, and hence more power dissipation. These implications require trade-offs between the output amplitude and power dissipation.

Spectral purity: The sidebands appear around the oscillation frequency. These sidebands are called "phase noise" in the frequency domain and "jitter" in the time domain. The low frequency noise from active devices of the amplifier and resistive loss of the resonator are phase modulated and appear as sidebands around the oscillation frequency. The phase noise is used to quantify the spectral purity of an oscillator in wireless applications.

Power dissipation: VCO is the most power consuming block in a PLL system. Low power VCO design involves trade-offs with almost all other VCO parameters.

Usually, the tuning range and phase noise are the most critical performance parameters, and they are usually specified along with a power budget.

2. PLL Frequency Synthesis

The most commonly used frequency synthesis techniques are the direct analog synthesis, the direct digital synthesis, and the indirect synthesis (PLL based) [14–16].

The direct analog synthesis and the direct digital synthesis are both hardware intensive techniques to achieve fine frequency resolution and fast switching times. However, both of these methods are not suited for high frequency and low phase noise frequency synthesis. The PLL based indirect synthesis technique has the capability to generate high frequency signals with low phase noise. The PLL based frequency synthesizers are used in most wireless applications for local oscillator generation. The PLL frequency synthesizers are considered in this work for LO signal generation.

Phase-Locked Loop (PLL) Overview

PLL frequency synthesizers are used for local oscillator (LO) signal generation in wireless radio transceivers. A typical architecture of a PLL synthesizer is shown in Figure 2.3 which is consist of a VCO, a low-pass loop filter, a phase/frequency detector (PFD), charge-pump (CP), and a frequency divider. A PLL forces the divided frequency to be exactly equal to the frequency of the input signal in the lock mode, i.e., $f_{ref} = f_{out}/N$. The divider value is digitally controlled to select the desired channel.

Modern high-performance integrated frequency synthesizers employ charge-pump PLL (CP-PLL) due to the fact that its frequency capture (pull-in) range is limited by the VCO tuning range and the CP output voltage. Another advantage of the CP-PLL is that the DC gain of the open-loop phase transfer function of the PLL is infinity, leading to a zero steady-state phase error [17]. In this section, CP-PLL design is over-viewed for passive loop filter implementation

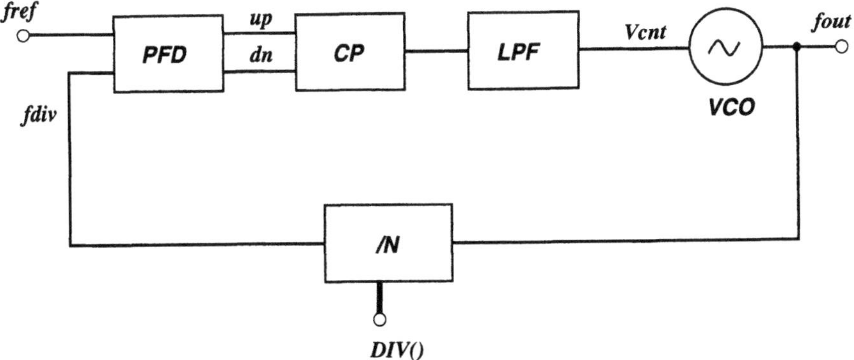

Figure 2.3. A typical phase-locked loop architecture.

using linear PLL model. In depth treatment of PLL frequency synthesis can be found in the references [14–16].

A phase-locked loop (PLL) is a simple negative feedback system which can generate multiplication of a reference signal. A PLL model is shown in Figure 2.4 as a negative feedback system, where

$$\frac{\theta_{out}(s)}{\theta_{ref}(s)} = T(s) = \frac{G(s)}{1 + G(s)H(s)} \tag{2.2}$$

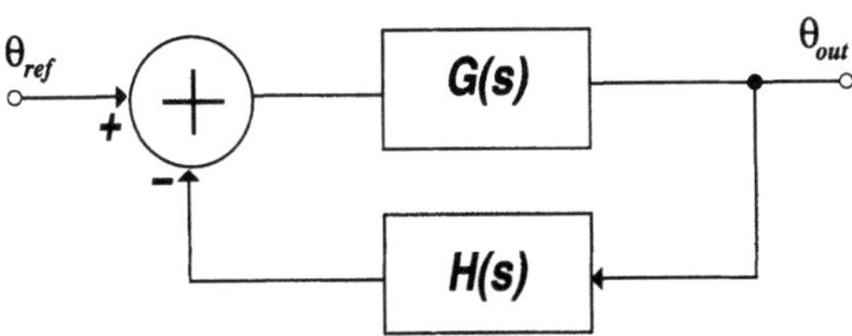

Figure 2.4. PLL as a negative feedback model.

A mathematical model for linear representation of the phase of a PLL behavior in the locked state is presented in Figure 2.5(a). Control theory is often utilized to characterize PLL behavior. The passive loop filter design for a type-II 3rd order charge pump PLL (CP-PLL) is overviewed here. The type of the system refers to the number of poles at the origin for the open loop transfer function, $G(s)H(s)$. The order of the system refers to the highest degree of the polynomial expression in the closed loop transfer function, $T(s)$.

From Fig. 2.5(a), the forward transfer function in the feedback system can be written as;

$$G(s) = \frac{I_{CP}}{2\pi} \cdot F(s) \cdot \frac{2\pi K_{VCO}}{s} \tag{2.3}$$

Where $I_{CP}/2\pi$ is the gain of PFD/CP, $2\pi K_{VCO}$ is the gain of VCO, and $F(s)$ is the transfer function of the low pass filter. The passive loop filter configuration for 3rd order PLL is shown in Fig. 2.5(b).

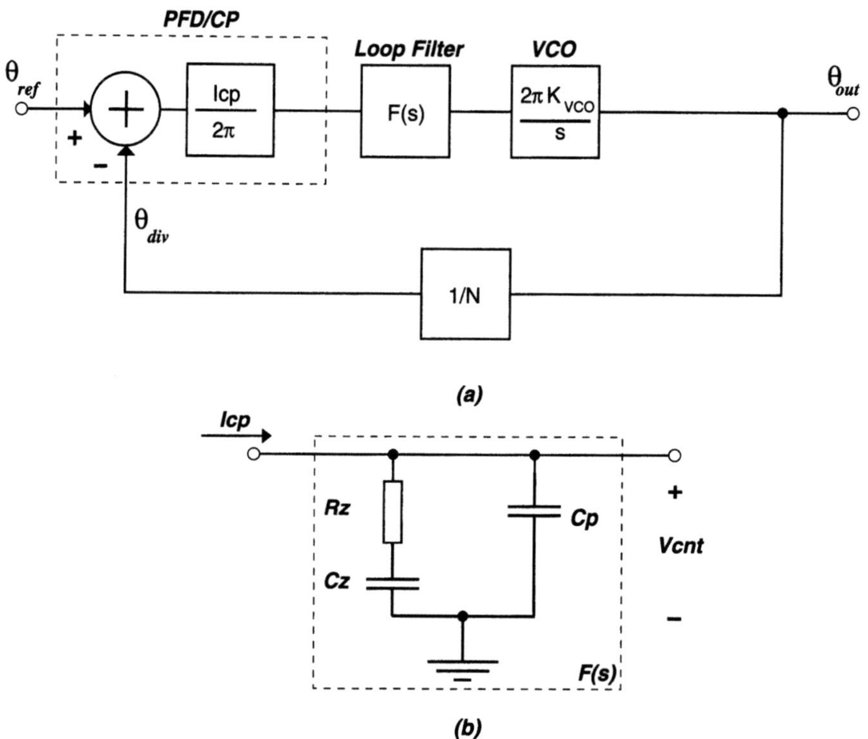

Figure 2.5.   (a) PLL linear model (b) Three element passive loop filter.

The transfer function of the loop filter, $F(s)$, for loop filter configuration in Figure 2.5(b) is be expressed as;

$$F(s) = \frac{V_{cnt}}{I_{CP}} = \frac{sR_zC_z + 1}{s(sR_zC_zC_p + C_z + C_p)} \tag{2.4}$$

The feedback transfer function is,

$$H(s) = 1/N \tag{2.5}$$

The open-loop transfer function is written from Equations 2.3 and 2.5,

$$G(s)H(s) = \frac{I_{CP}}{2\pi} \cdot F(s) \cdot \frac{2\pi K_{VCO}}{s} \cdot \frac{1}{N} \tag{2.6}$$

The closed-loop transfer function is written as,

$$\frac{\theta_{out}(s)}{\theta_{in}(s)} = \frac{G(s)}{1 + G(s)H(s)} \tag{2.7}$$

The open-loop frequency response, $G(s)H(s)\big|_{s=j\omega}$, is used to determine loop filter parameters by ensuring the loop stability [9, 15]. A Bode plot illustrates the asymptotic behavior of the magnitude and the phase of the open-loop transfer function according to the pole and zero locations. The bode plot of the open-loop transfer function is shown in Figure 2.6. The frequency, ω_c, where the open-loop gain drops to unity is called the gain cross-over frequency. The phase margin, ϕ_m, is defined as the difference between 180 degrees and the open-loop phase value at the cross-over frequency. The phase margin is usually chosen between 50 and 70 degrees.

The frequency response of the open-loop can be expressed in terms of the loop filter time constants $\tau_p = R_z \cdot \frac{C_z C_p}{C_z + C_p}$ and $\tau_z = R_z C_z$, and the loop parameters, I_{CP}, K_{VCO}, and N.

$$G(s)H(s)\big|_{s=j\omega} = -\left(\frac{I_{CP}K_{VCO}}{\omega^2 N(C_p + C_z)}\right) \cdot \left(\frac{1 + j\omega\tau_z}{1 + j\omega\tau_p}\right) \tag{2.8}$$

If the loop filter parameters are provided, the crossover frequency can be expressed as [9],

$$\omega_c = \frac{I_{CP}K_{VCO}R_z}{N} \cdot \frac{C_z}{C_Z + C_p} \tag{2.9}$$

If the zero, $\omega_z = 1/T_z$, is chosen a factor α below ω_c and the third pole, $\omega_p = 1/T_p$, is chosen a factor β above ω_c, the phase margin for loop stability is guaranteed [9]. A phase margin of 60° can be obtained by choosing the factors α and β as four [9]. The component values for loop filter can be calculated from loop parameters as following;

$$R_z = \frac{N}{I_{CP}K_{VCO}} \cdot \omega_c \tag{2.10}$$

$$C_z = \frac{I_{CP}K_{VCO}}{N} \cdot \frac{\alpha}{\omega_c^2} \tag{2.11}$$

$$C_p = \frac{I_{CP}K_{VCO}}{N} \cdot \frac{1}{\beta\omega_c^2} \tag{2.12}$$

Analysis for higher order passive loop filter and active loop filter designs can be found in the reference [8] [9] [18].

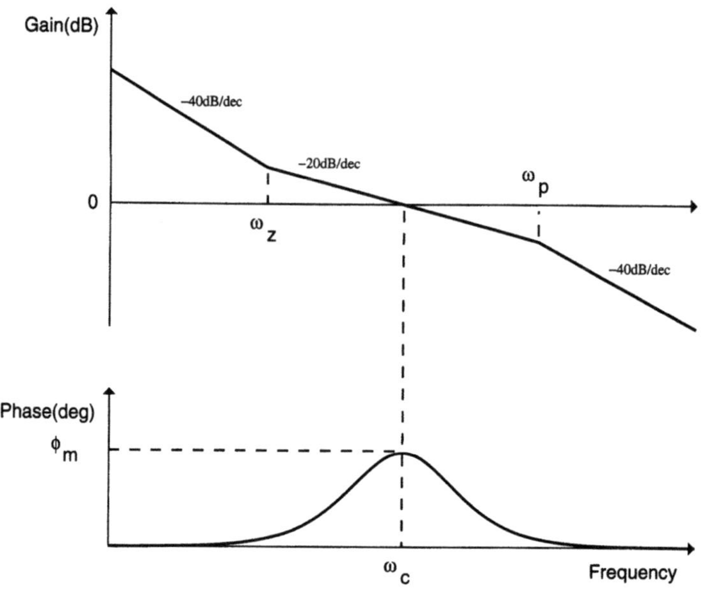

Figure 2.6.   PLL Bode plots.

PLL Dynamic Characteristics

The dynamic behavior of a PLL at startup or when changing division ratio
is one of the important design parameters of a PLL. The dynamic behavior of a
PLL involves two different steps; frequency acquisition and phase correction.
In the first step, the PLL starts with an unlocked condition at the power up
and achieves frequency lock to the channel frequency. In the second step,
the PLL is locked to the correct frequency and there is a small phase error,
$\theta_{err}(t) = \theta_{ref}(t) - \theta_{div}(t)$. We are interested in finding how long it takes
for a PLL output frequency to settle within a given frequency error, such as
$\pm 1KHz$. To find the response of PLL to a phase step at the input or (output),
the closed loop transfer function of the PLL is multiplied by a step function, and
the inverse Laplace Transform is taken to convert the response to a time domain
response. The closed loop transfer function of PLL as shown in Figure 2.5(a)
is expressed [8] as a second order system,

$$H(s) = \frac{\theta_{out}}{\theta_{in}} = \frac{2\zeta\omega_n s + \omega_n^2}{s^2 + 2\zeta\omega_n s + \omega_n^2} \tag{2.13}$$

By getting the inverse Laplace from Equation 2.13 and ignoring higher order
terms, the lock time for a given frequency jump at the output is expressed

as [8, 18]:

$$t_{lock} = \frac{-1}{\zeta \omega_n} \ln \left(\frac{f_{accuracy}}{f_{jump}} \sqrt{1 - \zeta^2} \right) \tag{2.14}$$

where the natural frequency of the second order system, ω_n, and the damping factor, ζ, are written in terms of the PLL loop parameters [16],

$$\omega_n = \sqrt{\frac{I_{CP} K_{VCO}}{N(C_p + C_z)}} \tag{2.15}$$

$$\zeta = \frac{R_z C_z}{2} \omega_n \tag{2.16}$$

The parameters, ω_n and ζ are useful for transient analysis and are based on a second-order model of the system. In practice, PLL frequency synthesizers are never a system of the second-order as there are usually additional poles to suppress reference spurs. Open-loop bandwidth and phase margin are useful concepts for analysis of frequency response and stability of PLL systems. Open-loop bandwidth and phase margin are closely related to the second order system parameters [18];

$$\omega_c = 2\zeta \omega_n \tag{2.17}$$

$$\sec \phi_m - \tan \phi_m = \frac{1}{4\zeta^2} \tag{2.18}$$

Equations 2.17 and 2.18 suggest that the lock time of a PLL is closely related to the PLL frequency response. Therefore, there is a trade-off between loop performance parameters and the lock time.

3.   Phase Noise Specification

The phase noise performance of an LO signal plays a key role for the overall performance of the wireless system by determining how closely channels can be placed in narrow-band systems and how closely constellation points can be placed in the I/Q plane in digital modulated systems. The first one specifies the phase noise power level with respect to carrier at a given offset frequency (dBc/Hz). This specification is usually important in narrow-band systems such as cellular systems. The cellular standards strictly define the radio performance parameters in great details for efficient use of the licensed RF bands for more users. The second one specifies the integrated phase noise performance over the signal bandwidth. The second specification is usually more important in broadband systems (WLAN) where high data rate is transmitted to a short distance using a wide bandwidth channel. These phase noise specifications are discussed in this section.

The phase noise specification of a frequency synthesizer at large offset frequency can be determined from standard specified blocker requirements for in-band (adjacent and non-adjacent channels) and out of band signals [11]. The undesired sideband energy from the LO phase noise reciprocal mixes with the in-band or out-of-band undesired signals which generates interfering signal, I, within the desired signal band. The power levels for blockers along with the desired signal and the required BER, which translates to a minimum C/I at the receiver output, are given by the RF standards. It is assumed that a system noise floor, N, which include receiver front end noise (from antenna to mixer output) is present at the mixer output. The interfering signal generated by reciprocal mixing of blocker with the phase noise at the mixer output will add up to the receiver noise in the desired band as shown in Figure 2.7. The total undesired signal level at the mixer output will be the sum of the front-end system noise and interfering signal caused by reciprocal mixing, $N + I$. If $(N + I)$ is set to a minimum for a receiver to meet the required BER, then the interfering signal (noise), I, should be below the N level. Leaving a 3dB margin, the minimum C/I can be written in terms of the minimum C/N as,

$$C/I_{min} \geq C/N_{min} + 3dB \qquad (2.19)$$

The phase noise specification for given blocker levels and C/I_{min} at specified frequency can be calculated using the equation,

$$L(f_{spec}) \leq P_d - P_b(f_{spec}) - 10 \log BW - C/I_{min} \qquad (2.20)$$

where,
$L(f_{spec})[dBc/Hz]$ is the phase noise specification at the frequency offset, f_{spec}.
$P_d[dBm]$ is the desired signal power level.
$P_b(f_{spec})[dBm]$ is the blocker level at the the frequency offset, f_{spec}.
$BW[Hz]$ is the signal bandwidth.
$C/I_{min}[dB]$ is the required minimum carrier-to-interference level.

As an example, if we consider the GSM mobile terminal blocking profile to calculate the phase noise specification, the desired signal power level is -102dBm, the blocker level at 600kHz offset frequency is -43 dBm, the channel bandwidth is 200kHz, and the required C/N_{min} is 9dB to meet the BER specification. Then, the minimum phase noise level at 600kHz offset frequency should be,

$$L(600KHz) = -102 - (-43) - 10 \log(200 \cdot 10^3) - (9 + 3) = -124dBc/Hz$$
$$(2.21)$$

The interferer power, P_I, is equal to $P_I = L(600kHz) + P_b(600kHz) + 10 \log BW = -124dBc/Hz - 43dBm + 53dB = -114dBm$. The thermal noise of receiver at the mixer output is equal to $P_N = P_W(= kT) +$

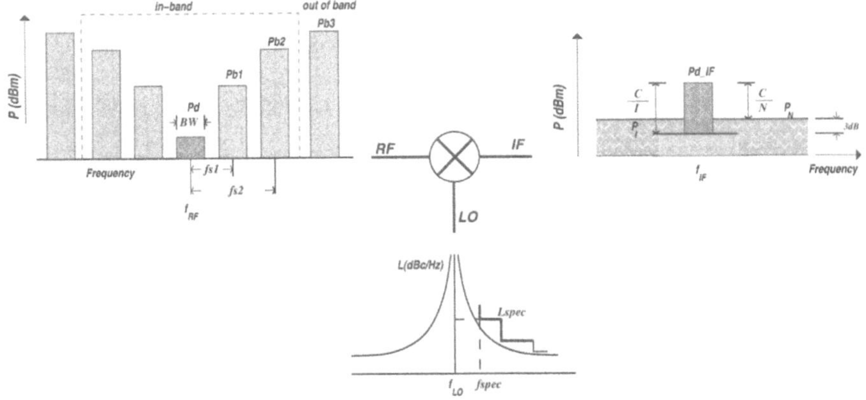

Figure 2.7.   LO phase noise specification

$10 \log BW + NF$. NF is the cascade noise figure of the receiver up to the mixer output from the LNA input. If the noise figure, NF, 10 dB, then the thermal noise power level at the mixer output is $P_N = -174dBm/Hz + 53dB + 10dB = -111dBm$.

It should be noted that blockers will degrade the receiver performance in three different mechanisms; reciprocal mixing (local oscillator phase noise), gain desensitization (compression of amplifier, 1-dB compression point), and inter-modulation (IM) distortion . Reciprocal mixing effect has been discussed previously. The gain desensitization and IM distortion have implications on the front-end linearity. Furthermore, there will be other interfering signals falling into the desired band in addition to the interferer caused by the blockers. The other interferer will come from image signals, second-order distortion (IP2), and spurious mixing products (2RF-3LO, etc.). The input IP2 performance is of particular significance in a direct-conversion (zero IF) receiver since second-order nonlinearity effects can down-convert continuous wave (CW) as well as AM modulated blockers to DC or near DC. The image signal is more of a concern for low IF and wide-band IF receiver architectures. The spurious mixing products appear in wide-band IF receivers. These all should be taken into consideration in the calculation receiver specification.

The phase noise of the LO is added to the phase of the RF signal in the up and down conversion processes. The phase noise exhibits residual phase deviation at the recovered baseband signal phase. The rms residual phase deviation of the LO signal in degrees is expressed as,

$$\text{Residual phase deviation (in degrees)} = \Phi_{res} = \frac{180}{\pi} \cdot \sqrt{\int_{f_l}^{f_h} \phi_{LO}^2(f)df}$$

(2.22)

where $\phi_{LO}(f_m)$ is the phase noise spectral density of the LO signal in $rad/\sqrt{Hz}$, and it is related to the single-side band (SSB) phase noise, $L(f_m)$ (in dBc/Hz) with,

$$\phi_{LO}(f_m) = \sqrt{2} \times 10^{L(f_m)/20} \qquad (2.23)$$

or

$$L(f_m) = 10\log\left(\frac{\phi_{LO}^2(f_m)}{2}\right) \qquad (2.24)$$

Equation 2.22 is expressed in terms of the SSB phase noise,

$$\text{Residual phase deviation (in degrees)} = \Phi_{res} = \frac{180}{\pi} \cdot \sqrt{\int_{f_l}^{f_h} L(f)df} \qquad (2.25)$$

Also, the integrated phase noise is also expressed,

$$\text{Integrated phase noise (in dBc)} = \int_{f_l}^{f_h} L(f)df \qquad (2.26)$$

4. Summary

An overview of the frequency synthesizer and oscillator in wireless radio transceiver is presented. The quality of the LO signal has significant impact on the performance of the radio transceiver. The VCO parameters are described. The PLL frequency synthesis and the PLL loop filter design are discussed. The phase noise is specified by considering blocker requirement and rms phase error for digital modulation schemes.

Chapter 3

PLL PHASE NOISE ANALYSIS

The VCO used in the traditional PLL frequency synthesizer is an external block which is either a module or built with high performance discrete components to meet the requirements with a fairly large margin. Therefore, the synthesizer employing an external VCO decouples the VCO design from the rest of the PLL design for phase noise consideration. Over designed for phase noise, an external VCO offers sufficient degrees of freedom to meet the PLL performance parameters independently. This is not the case for fully integrated frequency synthesizer with the VCO in CMOS technology. The VCO phase noise requirement can only be met with a small margin, and hence this strongly couples PLL and VCO design for phase noise consideration.

This chapter explores the phase noise characteristics of a PLL frequency synthesizer. First, the phase noise characteristics of the oscillator are investigated with low phase noise design in mind. An analytical model is developed to investigate the noise properties of the closed-loop PLL. The closed-loop PLL noise is an important factor of the total system performance in modern digital communications which use phase modulation, such as QPSK and QAM. The PLL noise model includes the effect of the blocks forming the PLL as well as the VCO. A SPICE implementation of the model is also compared with measured results.

1. Phase Noise Definition

The most critical specification for any oscillator is its spectral purity. The output spectrum of an ideal oscillator is an impulse at, ω_o, as shown in Fig. 3.1(a). An ideal oscillator can be described as a pure sine wave in the time domain,

$$V(t) = V_o \sin(\omega_o t) \tag{3.1}$$

However, in any practical oscillator, the spectrum has power distributed around the center frequency (ω_o) as shown in Fig. 3.1(b). In addition, the power is also distributed at the harmonics of the oscillation frequency, i.e., ($2\omega_o$, $3\omega_o$, $4\omega_o$, ...). The instantaneous output of a practical oscillator may be represented by;

$$V(t) = V_o[1 + A(t)]\sin(\omega_o t + \phi(t)) + higher\ order\ harmonics \quad (3.2)$$

where $A(t)$ and $\phi(t)$ represent amplitude and phase fluctuations of the signal, respectively. There are two types of phase terms appearing at the output spectrum of an oscillator. The first type appears as a distinct component in the spectrum, and it is referred to as a *spurious tone or signal* as shown in Fig. 3.1(b). The second type appears as random phase fluctuations, and it is referred to as *phase noise* as shown in Fig. 3.1(b). The phase noise in an oscillator is mainly due to internal noise sources such as thermal noise and active device noise sources (flicker noise or 1/f, shot noise). They are random in nature. The internal noise sources set a fundamental limit for a minimum obtainable phase noise in oscillator design.

The spurious tones are due to external noise sources such as noise on control voltage, power supply, and bias current coupled clock signals. They are deterministic in nature. The spurious tones are not directly related to the oscillator but are important in the frequency synthesizer output and the PLL design specifications.

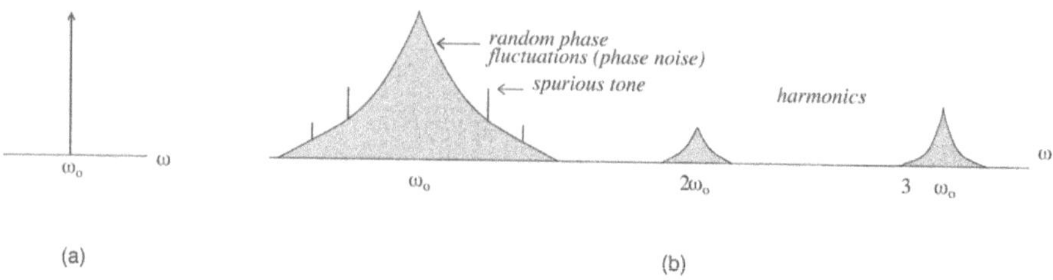

Figure 3.1. The oscillator output spectrum of an ideal oscillator (a) and a practical oscillator (b).

The phase noise of an oscillator is typically quantified by the single-sideband (SSB) phase noise, which is defined as the ratio of noise power in a 1 Hz bandwidth at an offset, f_m, to the signal power, as shown in Fig. 3.2. Single-sideband (SSB) phase noise is specified in dBc/Hz at a given frequency offset,

f_m, from the signal frequency.

$$L(f_m) = 10 \log\left(\frac{P_{NOISE}(f_m)}{P_{SIGNAL}}\right) \qquad (3.3)$$

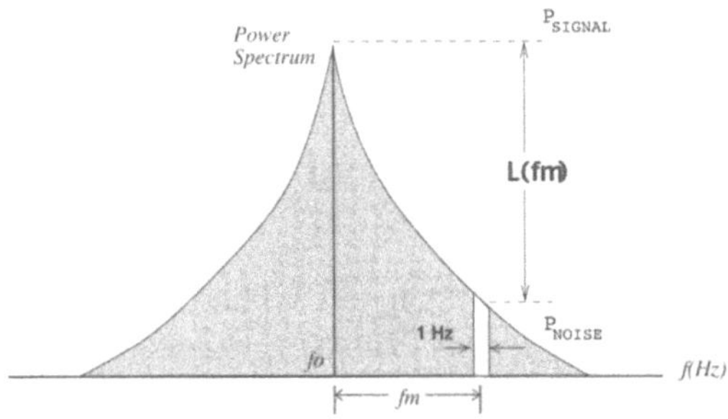

Figure 3.2. Phase noise definition.

Equation 3.3 is also can be written as,

$$L(f_m) = 10 \log\left(\frac{v_{n,rms}^2(f_m)}{V_{c,rms}^2}\right) \qquad (3.4)$$

where $v_{n,rms}^2(f_m)$ is the rms value of the sinusoid representing the phase noise sideband at the offset frequency f_m and $V_{c,rms}$ is the rms value of the carrier signal.

Sidebands around the carrier shown in the Figure 3.2 can be from PM or AM modulation of the carrier signal by noise. The phase noise in Equation 3.3 is written as AM and PM noise contributions as,

$$L(f_m) = 10 \log\left(\frac{S_\phi(f_m)}{2} + \frac{S_a(f_m)}{2}\right) \qquad (3.5)$$

where $S_\phi(f_m)$ is the double-sideband (DSB) phase noise spectral density and $S_a(f_m)$ is the DSB amplitude noise spectral density.

Consider a signal of constant amplitude A_c which is phase-modulated by a sine wave of frequency f_m,

$$s_{VCO}(t) = A_c \cos(\omega_c t + \theta_p \sin \omega_m t) \qquad (3.6)$$

where θ_p is the peak phase deviation, also known as the modulation index. When $\theta_p \ll \pi/2$, then the narrow band FM approximation can be used to obtain [14],

$$s_{VCO}(t) = A_c(\cos(\omega_c t) - \frac{\theta_p}{2}\cos(\omega_c - \omega_m)t + \frac{\theta_p}{2}\cos(\omega_c + \omega_m)t) \quad (3.7)$$

The amplitude of the side band, A_{sp}, due to modulating signal can be written as,

$$A_{sp} = \frac{A_c\theta_p}{2} \quad (3.8)$$

Also, from the above equation, the peak phase deviation, θ_p, due to a single sideband tone with amplitude, A_{sp}, around the carrier is expressed as,

$$\theta_p = \frac{2A_{sp}}{A_c} \quad (3.9)$$

Assuming that the side bands are due to the phase noise fluctuations, then the phase noise power spectral density can be written using Equations 3.4 and 3.9 as;

$$\frac{S_\phi(f_m)}{2} = \frac{\phi^2_{rms}(f_m)}{2} = \left(\frac{v_{n,rms}(f_m)}{V_{c,rms}}\right)^2 = 10^{L(f_m)/10} \; [rad^2/Hz] \quad (3.10)$$

or,

$$L(f_m) = 10\log\left(\frac{\phi^2(f_m)}{2}\right) \quad (3.11)$$

Equations 3.10 and 3.11 are of fundamental importance for the treatment, calculation and simulation of phase noise in PLLs. While $L(f_m)$ is useful to measure and characterize with measurement tools, $\phi(f_m)$ is useful in calculations of integrated residual phase deviation over a given bandwidth [8].

The effect of AM modulation of a carrier signal is considered. An AM modulated signal can be written as,

$$s_{VCO}(t) = A_c(1 + m\cos\omega_m t)\cos\omega_c t \quad (3.12)$$

where m is the AM modulation index, f_m is the modulation frequency. Equation 3.12 can be expanded as;

$$s_{VCO}(t) = A_c[\cos\omega_c t + \frac{m}{2}\cos(\omega_c + \omega_m)t + \frac{m}{2}\cos(\omega_c - \omega_m)t] \quad (3.13)$$

Equation 3.13 indicates that AM modulation generates a pair of spurious sidebands similar to those of PM modulation (narrow-band). Only difference between the sidebands of AM and PM is the phase relationship as shown in Figure 3.3. When $L(f_m)$ is measured by a spectrum analyzer, it does not distinguish the AM and PM sidebands since phase information is not retained by a spectrum analyzer. Close to the carrier, PM modulated noise dominates, and can be considered as phase noise. Far offset frequencies, both AM and PM noise contribute equally [14].

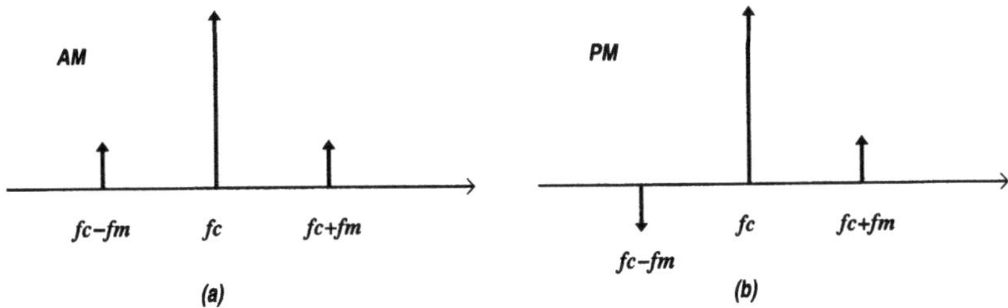

Figure 3.3. Sidebands around carrier due to (a) AM (b) PM.

2. Oscillator Noise Characteristics

Noise characteristics of oscillators have a great impact on the overall noise characteristics of a PLL since each PLL typically employs two oscillators; a reference crystal oscillator and an RF VCO. Phase noise of a crystal oscillator has an impact on the close-in phase noise of a PLL while VCO noise shapes phase noise of a PLL at large offset frequencies. Effects of individual noise contribution of crystal and voltage controlled oscillators on the output PLL noise will be discussed in the PLL noise analysis section. Phase noise characteristics, modeling and simulation of oscillators are overviewed in this section.

An oscillator noise profile exhibits different slopes at various regions when the SSB phase noise is plotted in decibels versus log frequency as shown in Figure 3.4. These regions have slopes of 0, $1/f$ (-10 dB/decade), $1/f^2$ (-20 dB/decade) and $1/f^3$ (-30 dB/decade). The noise plot of an active device (MOS or Bipolar transistor) is shown in Figure 3.5 which has only two regions; thermal noise and $1/f$ region. The $1/f$ noise corner is in the vicinity of 500KHz to 1MHz for sub-micron CMOS technology [11], and it is in the vicinity of 1KHz to 10KHz for bipolar transistor.

The noise from the active and the passive devices are injected into the resonator of an oscillator. The injected noise is shaped by the VCO resonator which has -20 dB/decade slope on either side of the oscillation frequency. The flat thermal noise becomes $1/f^2$ in nature and the $1/f$ noise becomes $1/f^3$ in nature, within the resonator bandwidth, at the output of the oscillator (Figure 3.4(a)). Depending on the relative position of the 1/f corner and the resonator bandwidth, $f_o/2Q$, two cases arise as shown in Figure 3.4(a) and (b). Figure 3.4(a) is typical characteristic of the most RF oscillators with LC resonator tanks. Figure 3.4(b) is typical characteristic of a high-Q crystal oscillator.

Oscillator Noise Models and Simulation

Phase noise in oscillators has long been the subject of theoretical and experimental investigation. An early model of phase noise, introduced by Leeson [19],

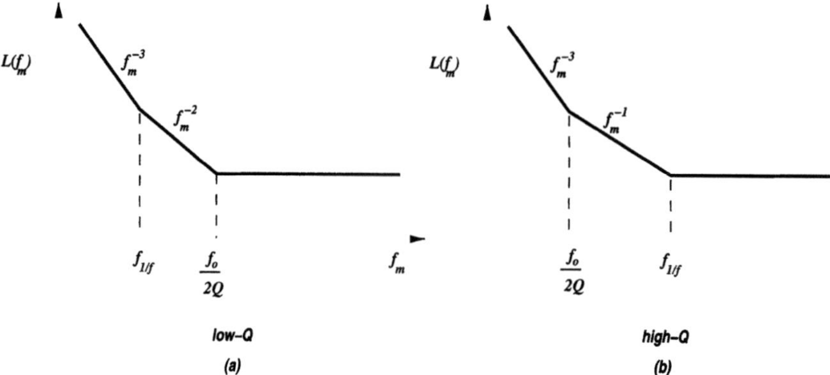

Figure 3.4. VCO phase noise characteristics (a) low-Q case (b) high-Q.

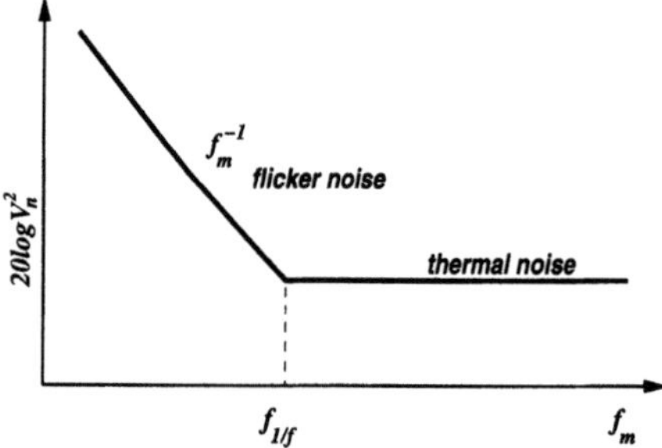

Figure 3.5. Noise characteristics of a MOS transistor at a fixed bias condition.

qualitatively described phase noise spectra in a variety of oscillators [20]. The derivation of the Leeson phase noise model is based on a linear time-invariant (LTI) approach to the analysis of noise in oscillators. The Leeson phase noise model is given as,

$$L(f_m) = 10 \log \left[\frac{2FkT}{P_s} \left(\frac{f_o}{2Q_L f_m} \right)^2 \left(1 + \frac{f_k}{f_m} \right) \right] \qquad (3.14)$$

where Q_L is the loaded quality factor of the resonator, f_o is the oscillation frequency, P_s is the signal power of the oscillator, f_m is the frequency offset, F is the noise factor for the active devices, k is the Boltzmann constant, T is the temperature in Kelvin, and f_k is the flicker noise corner frequency knee in the phase noise. f_k does not necessarily coincide with the flicker noise corner frequency of the active device.

Equation 3.14 shows clearly that three parameters Q, F, and P_s have significant effect on the phase noise of an oscillator. A low noise oscillator requires minimum F and maximum Q and P_s. Minimization of the noise factor F involves mostly active devices. MOS transistors have higher flicker noise corner frequency (from 500KHz up-to a couple of MHz) than a typical bipolar transistor. Output power, P_s should be maximized within a certain power budget. The loaded Q of the LC-tank for fully integrated RF VCO is limited by the available integrated spiral inductor and varactor in a given technology.

Phase noise models based on a linear time-invariant analysis provide insights and quantitative understanding of noise in oscillators and serve as a starting point in design procedure. An accurate estimation of the phase noise of an oscillator requires nonlinear simulations including all available noise models.

Mathematically more rigorous solutions by Hajimiri [21] and Demir [22] have been introduced in recent years since Leeson's approximate model. However, these solutions are mathematically involved, do not provide much intuition into design and analysis of circuits, and are not practical for hand calculations. These solutions are more appropriate for simulator implementation and verification.

SPICE small-signal noise simulation (.AC) does not provide adequate modeling of an RF oscillator noise. The state of art RF simulators provide a steady-state noise analysis which is based on the steady-state behavior of the circuit, rather than its DC operating point (EldoRF[23], SpectreRF [24], Agilent EEsof [25]). The circuit is linearized around the large-signal periodic time-varying operating point, and the noise contributions (the amplitudes of the noise signals) are assumed to be small enough so that the small-signal assumption is valid. Close to carrier (offset frequencies $1kHz < f$), phase noise simulations may not give accurate results because small-signal assumptions are not valid close to carrier (power level in the order of -30dBm and higher). However, it is often not very critical to get very accurate result for the VCO phase noise at very close offset frequencies for two reasons; first, most wireless standards modulation schemes do not carry information at very close offset frequencies to the carrier and second the VCO is placed in the PLL which suppresses the VCO noise within the loop bandwidth close the carrier.

3. PLL Noise Analysis

Phase noise of a VCO placed inside a PLL is shaped by the PLL noise transfer functions. A free running VCO phase noise is simply called VCO phase noise, while the phase noise of a VCO inside a locked PLL is called PLL output phase noise. The overall PLL output phase noise is characterized by the noise contributions of all blocks in a PLL.

A linear phase-domain model of a PLL with additive noise sources is shown in Figure 3.6. θ_{ref} represents the noise in $rad/\sqrt{Hz}$ that appears at the reference input to the PFD. It includes the crystal oscillator, crystal buffer, and reference divider noises. θ_{div} is the noise due to the divider in $rad/\sqrt{Hz}$. θ_{VCO} is the phase noise of the VCO in $rad/\sqrt{Hz}$. θ_{PFD} is the noise of PFD in $rad/\sqrt{Hz}$.

$v_{n,cnt}$ is the noise at the VCO control voltage input due to the loop filter and other coupled noise sources to the control line in $V/\sqrt{Hz}$. $i_{n,cp}$ is the noise of the CP current in $A/\sqrt{Hz}$. The transfer functions from these noise sources to the output can be written as;

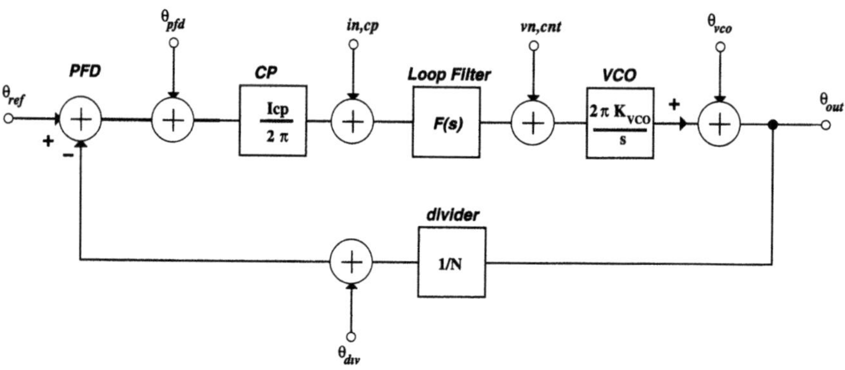

Figure 3.6.   PLL noise model.

$$\frac{\theta_{out}(s)}{\theta_{ref}(s)} = T_{ref}(s) = \frac{G(s)}{1 + G(s)H(s)} \tag{3.15}$$

$$\frac{\theta_{out}(s)}{\theta_{div}(s)} = T_{div}(s) = \frac{G(s)}{1 + G(s)H(s)} \tag{3.16}$$

$$\frac{\theta_{out}(s)}{\theta_{PFD}(s)} = T_{pfd}(s) = \frac{G(s)}{1 + G(s)H(s)} \tag{3.17}$$

$$\frac{\theta_{out}(s)}{i_{n,cp}(s)} = T_{cp}(s) = \frac{2\pi}{I_{cp}} \cdot \frac{G(s)}{1 + G(s)H(s)} \tag{3.18}$$

$$\frac{\theta_{out}(s)}{\theta_{vco}(s)} = T_{vco}(s) = \frac{1}{1 + G(s)H(s)} \tag{3.19}$$

$$\frac{\theta_{out}(s)}{v_{n,cnt}(s)} = T_{vcnt}(s) = \frac{2\pi K_{VCO}}{s} \cdot \frac{1}{1 + G(s)H(s)} \tag{3.20}$$

The transfer functions, $G(s)$ and $H(s)$, are defined as (from Figure 3.6);

$$G(s) = \frac{I_{cp}}{2\pi} \cdot F(s) \cdot \frac{2\pi K_{VCO}}{s} \tag{3.21}$$

$$H(s) = \frac{1}{N} \tag{3.22}$$

where I_{cp} is the CP current in, A. K_{VCO} is the VCO gain in Hz/V. $F(s)$ is the loop filter transfer function.

The noise transfer functions are helpful to understand the effect of individual block contributions to the PLL output noise. The total noise can be calculated by adding all the block noise contributions in an RMS sum,

$$
\begin{aligned}
\theta_{out}^2 &= (\theta_{ref} T_{ref})^2 + (\theta_{div} T_{div})^2 + (\theta_{PFD} T_{pfd})^2 + (i_{n,cp} T_{cp})^2 \\
&+ (\theta_{VCO} T_{vco})^2 + (v_{n,cnt} T_{vcnt})^2
\end{aligned}
\tag{3.23}
$$

or

$$
\begin{aligned}
\theta_{out}^2 &= \left(\frac{G(s)}{1+G(s)H(s)}\right)^2 \cdot \left[\theta_{ref}^2 + \theta_{div}^2 + \theta_{pfd}^2 + \left(i_{n,cp} \cdot \frac{2\pi}{I_{cp}}\right)^2\right] \\
&+ \left(\frac{1}{1+G(s)H(s)}\right)^2 \cdot \left[\theta_{vco}^2 + \left(v_{n,cnt} \cdot \frac{2\pi K_{VCO}}{s}\right)^2\right]
\end{aligned}
\tag{3.24}
$$

Analysis of the noise transfer function reveals more information about the PLL noise shaping effect. The common factor for reference, divider, PFD and CP noises is,

$$\frac{G(s)}{1+G(s)H(s)} \tag{3.25}$$

The amplitude response of this common factor is shown in Figure 3.7(a). It exhibits low pass characteristics, and at low frequencies, the noise of the PLL is dominantly contributed by the reference, divider, PFD and CP noises. Also, the noise contributions from reference, divider, PFD and CP are increased by the feedback divider ratio N inside the loop bandwidth.

The amplitude response of the common factor, $1/[1+G(s)H(s)]$, for VCO and the control line noises is shown in Figure 3.7(b) and, it exhibit high-pass filter behavior. The VCO and the control line noises are suppressed inside the loop bandwidth, and the VCO noise dominates beyond the loop bandwidth.

The PLL bandwidth determines the total PLL noise shape. For a minimum residual phase error at the PLL output (integrated noise), the PLL loop bandwidth is chosen at the intersection of close-in noise and VCO noise as shown in Figure 3.8(a) [8]. Performance loss occurs if a non-optimum bandwidth is used as shown in Figure 3.8(b) and (c).

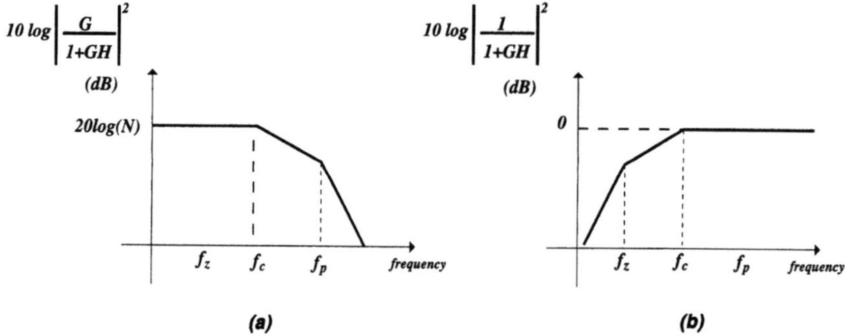

Figure 3.7. (a) transfer function for reference. divider, PFD, and CP noises (b) transfer function for the VCO and the control line noises.

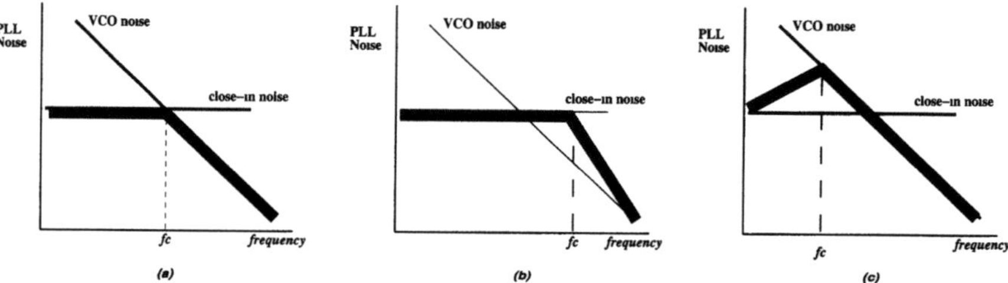

Figure 3.8. PLL bandwidth selection (a) optimum (b) too large (c) too narrow.

A typical SSB phase noise of a PLL is shown in Figure 3.9 for a well designed PLL. The phase noise behavior exhibits three different regions. The first phase noise region is the reference oscillator noise (about 100Hz-1KHz). The second region, also called PLL close-in noise, is dominantly contributed by PFD, CP, and the divider noise. The third region beyond the loop bandwidth is practically the VCO noise. The first and third regions are primarily determined by the oscillators; crystal and voltage-controlled. The second region, PLL close-in noise, is the indicator of the performance of a given PLL.

From Figures 3.7(a) and 3.9, we can define a PLL noise floor,

$$L_o = L_{pll,nf} + 20\log(N) + 10\log(f_{ref}) \qquad (3.26)$$

where L_o is the PLL close-in phase noise in dBc. $L_{pll,nf}$ is the noise floor of PLL (noise due to PLL circuitries; PFD, divider and CP). $20\log(N)$ is the increase of phase noise from the frequency magnification due to the feedback ratio, $1/N$. $10\log(f_{ref})$ is the increase of noise due to the reference frequency. When the reference frequency value increases, the CP starts to inject more noise into the loop filter. This will be further explained in the CP noise characteristics section.

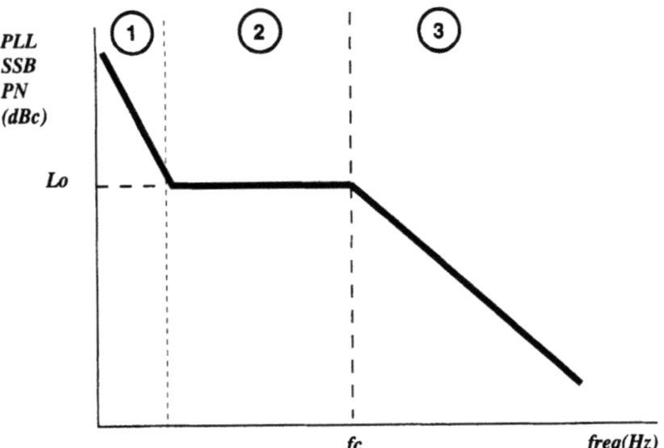

Figure 3.9. Typical PLL output phase noise.

For a measured PLL close-in noise, the noise floor of the PLL is found from,

$$L_{pll,nf} = L_o - 20\log(N) - 10\log(f_{ref}) \qquad (3.27)$$

Equation 3.27 provides a figure of merit for the PLL Synthesizer circuit itself. This allows comparison of different PLLs' performance by accounting for the feedback divide ratio, N and the reference frequency. Also, the noise floor of a PLL can be used to quickly identify the performance of a given PLL once N and the reference frequency are known.

Guidelines to optimize the output phase noise of a PLL frequency synthesizer are listed below,

- To minimize close-in phase noise (noise inside the loop bandwidth of PLL); reduce divider, CP, and PFD noises.

- To minimize close-in phase noise (noise inside the loop bandwidth of PLL); minimize divide ratio (N). The selection of divide ratio depends on the frequency synthesizer architecture. The minimum divide ratio N is limited by the output frequency resolution (channel bandwidth) in integer-N PLL architectures. Fractional-N architectures may be used to minimize divide ratio N [26].

- To minimize close in phase noise (noise inside the loop bandwidth of PLL) if dominated by CP noise; increase the CP current I_{cp} while keeping CP noise at the same level.

- To minimize phase noise outside the loop bandwidth of PLL; minimize VCO noise, minimize noise on the control line and minimize VCO gain K_{VCO}.

PLL Noise Modeling and Simulation

The closed-loop noise of a PLL can be calculated by using the analytical models developed in Section 3. Equation 3.23 can be implemented by using MATLAB, MATHCAD, EXCELL or with similar software. An implementation of the linear PLL noise model in SPICE is developed for closed-loop PLL noise simulation by representing phase variables as voltages and converting phase noise into voltage noise. The SPICE model is chosen because of the flexibility to include effects such as layout parasitics, bond-pads or active loop-filter components. The developed linear model is shown in Figure 3.10. This model includes the noise contribution of all PLL building-blocks. The CP and loop filter noises are represented in voltage/current since they are treated as current/voltage variable in the original loop. The reference, PFD, divider, prescaler, and VCO noises are converted from the SSB representation (dBc/Hz) to phase noise power spectral density (rad^2/Hz) using Equation 3.10.

$$\theta_{rms}^2(f_m) = 2(10^{L(f_m)/10}) \quad [rad^2/Hz] \tag{3.28}$$

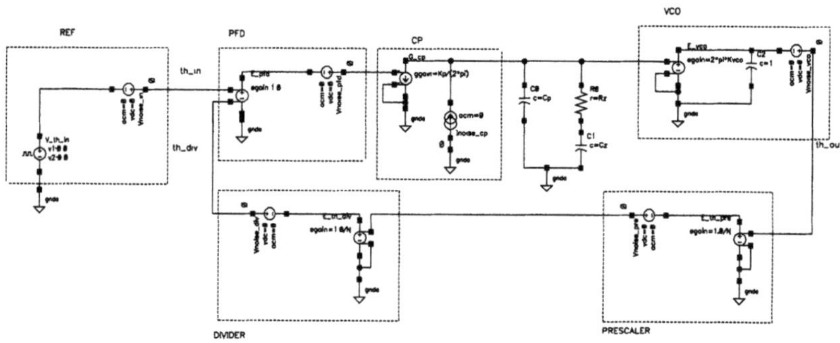

Figure 3.10.   Developed linear PLL model for noise calculation

The same model is also used to predict PLL stability and transient behavior. The model allows the prediction of noise contribution from individual PLL blocks, and hence it can be used for optimization of the overall PLL noise performance. An optimized PLL noise performance leads to relaxed VCO noise requirements by avoiding to overdesign.

Steps to Calculate Closed-loop Noise of a PLL

The major steps to calculate closed-loop noise of a PLL using the developed model are;

- Determine loop filter parameters for a given phase margin and loop bandwidth along loop parameters (VCO gain, charge-pump gain, divide ratio, reference frequency).

- Determine noise characteristics of individual PLL blocks by simulation or measurement.

- Use the individual noise characteristics in the model to compute PLL output noise.

Noise Characteristics of Individual PLL Blocks

Noise characteristics of individual PLL blocks must be extracted for PLL noise simulation using the developed model. The noise characteristics of individual PLL blocks can be extracted through measurement or/and simulation. The crystal oscillator phase noise characteristics under certain operation conditions are given by manufactures. Also, it can be measured since it is off-chip. The VCO noise can be characterized by measurement or simulation. The noise characteristics of oscillators are discussed in Section 2.0.

However, it is usually difficult to extract phase noise information for the other PLL blocks (divider, PFD, CP) from measurement. This is due to difficulty of a measurement setup for these blocks, and the noise levels of these blocks are below most of the measurement instruments' noise floors requiring special expensive instruments. Simulation is more convenient to extract noise information for these blocks.

Simulation of divider noise is straightforward. A pure single tone is applied at the divider input, and the SSB noise is measured at the divider output. Figure 3.11 shows the simulation setup and the simulated phase noise of a divide-by-8 prescaler.

A typical PFD and CP configuration is shown in Figure 3.12. The contribution of PFD noise is twice the noise of one flip-flop since both the flip-flops in PFD contribute equally to the phase noise. The phase noise of one flip-flop in PFD is simulated and used in the simulation model, and it is multiplied by 2 in the model.

An ideal CP-PLL with zero phase error neither sources current to, nor sinks current from, the loop filter. However, PLLs with zero phase error are insensitive to small loop-phase deviations due to finite signal rise times in the PFD and charge pump which is also called the "dead-zone" [17]. A commonly employed solution to "dead-zone" problem is to use an artificial phase offset so that CP pumps/sinks current when PLL is locked. When the PLL is locked, the average output current flowing into the loop filter as shown in Figure 3.12 is zero. Both the UP and DN currents are on for the duration of the "dead-zone" pulse. Even though the average current is zero, noise is injected from both UP and the DN currents for the duration of the "dead-zone" pulse. The charge pump current

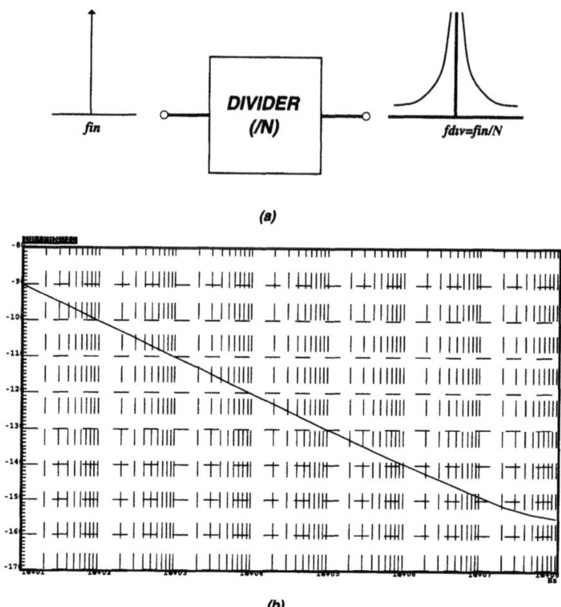

Figure 3.11.   (a) Divider noise simulation setup (b) Simulated SSB noise of divide-by-8 prescaler circuit.

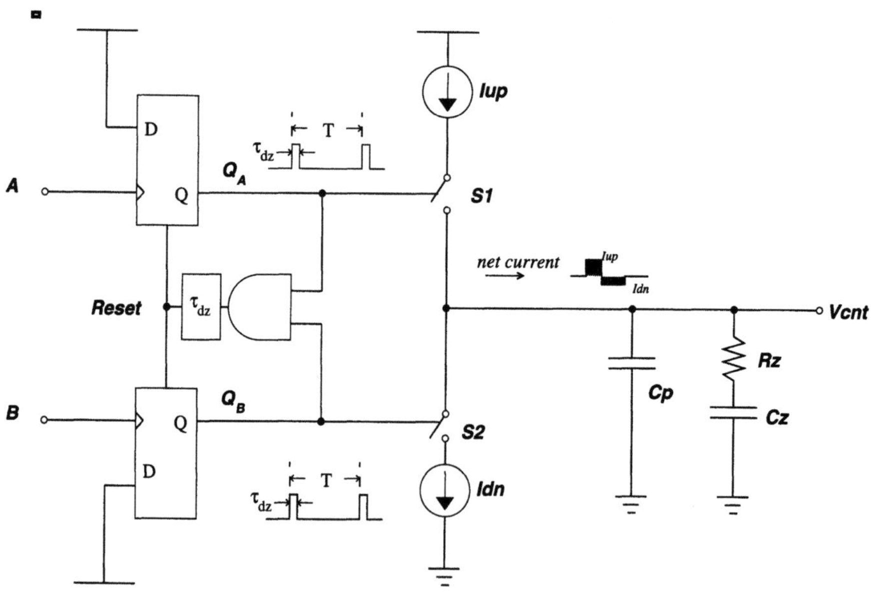

Figure 3.12.   PFD with charge pump.

noise injected to the loop filter under lock condition can be calculated as,

$$i_{n,cp}^2 = 2 \left(\frac{\tau_{dz}}{T} \right) I_{CP,noise}^2 \tag{3.29}$$

where factor 2 is used to account for UP and DN current pulses. $I_{CP,noise}$ is the current noise of the CP in $A/\sqrt{Hz}$. τ_{dz} is the dead-zone pulse width. T is the period of the reference signal.

Equation 3.29 suggests that if the reference frequency is increased, the CP will pump more noise into the loop filter. Hence, the PLL close-in noise given by Equation 3.26 will increase with the reference frequency.

The practical implementation of charge-pump circuit presents non-ideal effects; (i) the leakage currents (ii) the mismatch between magnitudes of up and down currents (iii) switching time mismatches between the up and down pulses [17]. These non-ideal effects are important for the reference spur calculation. They do not have significant effect on the noise calculation presented here if the CP circuit is properly designed. Equation 3.29 is sufficient for the CP noise calculation. Two key parameters to reduce the CP noise contribution to the PLL output noise are the dead-zone pulse width and the CP current noise. The dead-zone pulse width and the CP current noise should be minimized to minimize the CP noise contribution.

Experimental Results and Noise Optimization Techniques

Several PLL circuits have been analyzed using the developed SPICE model. Simulated and measured results for a 4GHz PLL [27] are presented. The 4GHz PLL is designed for 2.4GHz/5GHz multi-band WLAN radio applications. A prototype is designed as a test structure. The 4GHz PLL is fully integrated including loop filter components in $0.18\mu m$ CMOS technology. Only required external sources are a reference clock signal, 1.8V supply voltage, and bias currents for individual blocks.

The designed PLL loop parameters are $I_{CP} = 30\mu A$, $N = 192$, $K_{VCO} = 100MHz/V$, $PM = 53^o$, and $\omega_c = 2\pi 200KHz$. The corresponding loop filter parameters are $R_z = 94k\Omega$, $C_z = 25.6pF$, and $C_p = 3.2pF$. The open-loop behavior is simulated by using the developed model by opening loop at the CP output. The open-loop behavior is shown in Figure 3.13. However, the measured PLL open-loop behavior is different than the inital design because the measured VCO gain is higher than the inital design. The measured VCO gain is about $400MHz/V$. The open-loop behavior is re-simulated with the measured VCO gain as shown in Figure 3.14. The loop bandwidth is moved to $593KHz$, and the phase margin is reduced to 39°. This change is verified with measurement as shown in Figure 3.15. The measured loop bandwidth is about $600KHz$. The peaking at the loop bandwidth occurs due to reduced phase margin. Figure 3.15 also shows both the simulated phase noise and measured phase noises.

The closed-loop noise is simulated using measured phase noise for VCO and external reference signal, and simulated noise for other blocks in the developed SPICE model. The individual noise contributions at the PLL output from the PLL blocks are shown in Figure 3.16. The close-in noise of PLL is dominated by the CP noise. The close-in phase noise should improve by increasing the CP current as suggested by the analytical models since the CP noise is dominating close-in PLL noise. This is confirmed with further testing. Figure 3.17 shows the measured and simulated PLL output noises for the CP current, $I_{CP} = 90\mu A$. The close-in PLL noise improves as expected. Also, more peaking around the loop bandwidth is observed for the CP current, $I_{CP} = 90\mu A$. This is because of the further reduction of the phase margin. The open-loop behavior for $I_{CP} = 90\mu A$ is shown in Figure 3.18. The loop bandwidth is moved to $1150KHz$, and the phase margin is reduced to 24°. The measured loop bandwidth in Figure 3.17 confirms the open-loop simulation.

The developed model is able to predict the total PLL noise within $5dB$ in this example. An important conclusion from this example is that the fully integrated loop filter components do not allow the optimization and evaluation of a PLL performance. Optimization of PLL close-in noise involves iteration of CP current value for minimum noise while keeping the loop characteristics constant. This requires adjustment of the loop filter component values when the CP current value is changed. Also, off-chip loop filter components allows characterization of the VCO noise from measurement with a locked PLL by setting the loop bandwidth very narrow ($< 1KHz$). Variation in loop parameters has significant impact in the PLL performance.

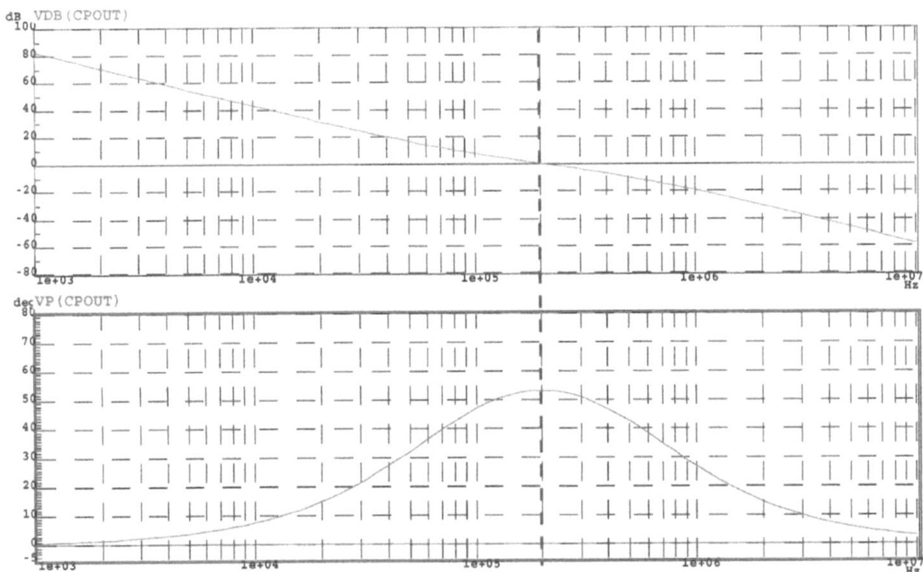

Figure 3.13. The intended open-loop behavior of the PLL.

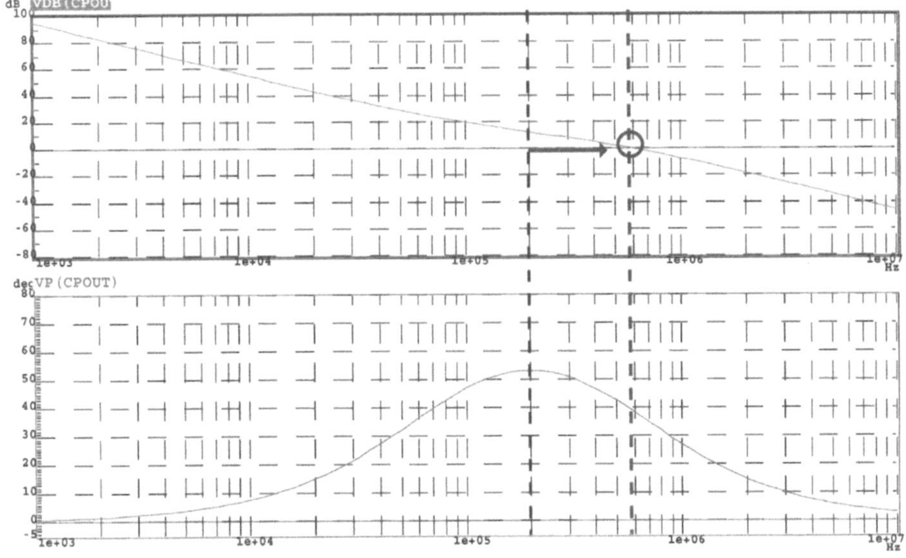

Figure 3.14. The actual open-loop behavior of the PLL due to high VCO gain.

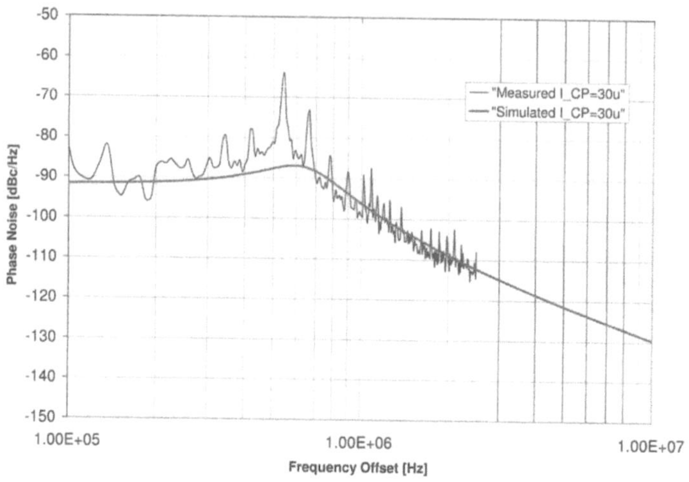

Figure 3.15. Comparison of simulated and measured output phase noise of 4GHz PLL at 3.84GHz for $I_{CP} = 30\mu A$.

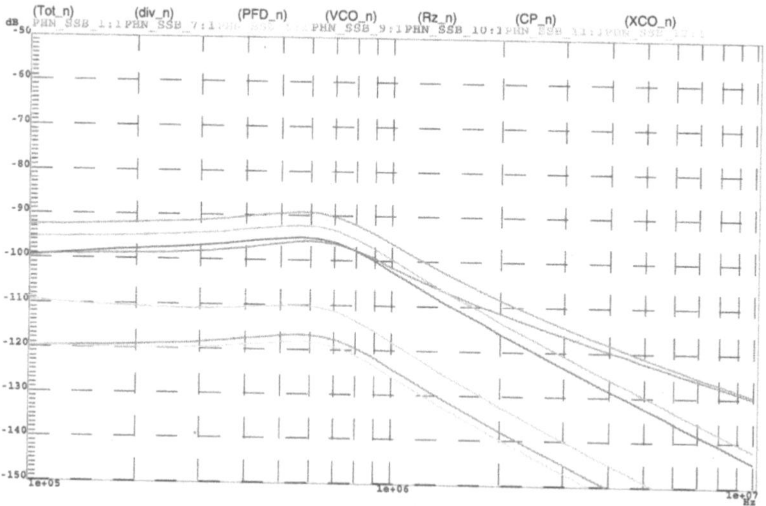

Figure 3.16. The simulated total phase noise of 4GHz PLL at 3.84GHz along with noise contributions from the PLL blocks.

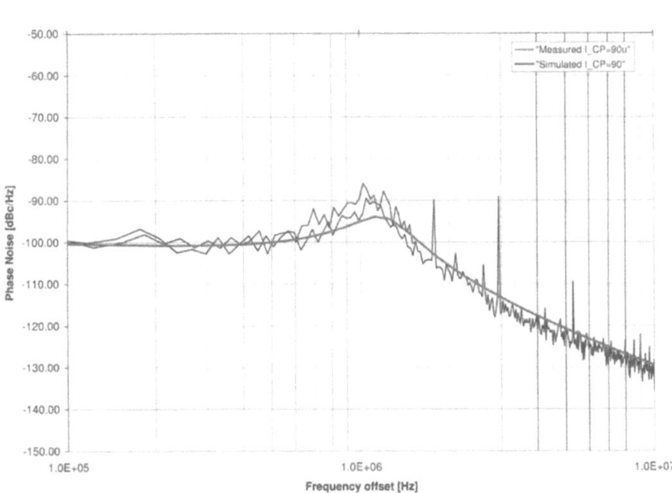

Figure 3.17. Comparison of simulated and measured output phase noise of 4GHz PLL at 3.84GHz for $I_{CP} = 90u$.

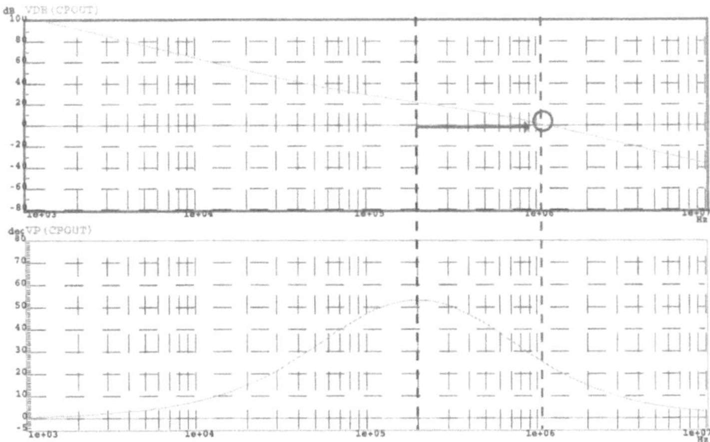

Figure 3.18. The open-loop behavior of the PLL for $I_{CP} = 90\mu A$.

4.   Summary

The analysis, calculation and optimization of the PLL output noise are presented in this chapter. Oscillator noise characteristics have important impact on the PLL phase noise since each PLL frequency synthesizer employs two oscillators; one high performance reference crystal oscillator and RF VCO. The calculation of the PLL output noise requires all individual PLL block noise characteristics.

Chapter 4

BROADBAND VCOs:
SYSTEM DESIGN CONSIDERATIONS

Modern multi-band/standard mobile radio transceivers may require to use multiple voltage control oscillator (VCO) functions to accommodate the many down/up conversions. Using separate VCOs would cause a substantial increase in cost and size of the radio frequency (RF) section. This component overload can be resolved by an integrated radio transceiver with a smart frequency planning and frequency synthesizer with a broadband VCO. The examples of high performance RF CMOS VCO are presented in recent years [28, 29].

The system design considerations for Broadband RF CMOS VCO in fully integrated transceiver solutions are explored in this chapter. As an individual block, VCO specifications include phase noise, tuning range and power consumption. However, full integration of an RF VCO in a transceiver also demands manufacturability, i.e. robust design against process, voltage, temperature (PVT) variation and tolerability to integrated environment disturbances such as substrate coupled noise, cross talk between signal lines, and supply line bounces. While the block specifications (phase noise, tuning range and power consumption) of a RF VCO have implications on the active circuit topology and resonator design, the manufacturability under PVT variation and the tolerability to integrated environment disturbances have implications on the system level (architecture, frequency planning, synthesizer architecture). First, top level architecture considerations issues are explored: radio architecture and frequency planning, frequency synthesizer architecture. Integration issues in CMOS technology are investigated.

1. Radio Architecture and Frequency Planning Considerations

Implementing a multi-band/standard radio transceiver requires an optimum frequency plan to maximize component sharing and minimize die area. Fully integrated receivers often employ either zero-IF (or low-IF depending on modulation scheme) and wideband-IF. Figures 4.1 and 4.2 depict the examples of zero-IF (or low-IF) and wide-band IF receiver architectures utilizing broadband VCOs for multi-standard applications where divide-by-2 and divide-by-4 circuits are used to generate quadrature LO signals.

Operating the VCO at a frequency that is higher than the required LO frequency has benefits for fully integrated direct-conversion radio implementation. It relaxes some of the direct-conversion architecture design issues such as excessive frequency pulling of the VCO by load pulling and injection locking, and self-mixing of RF and LO signals [30, 31]. Quadrature, I/Q, LO signals can be generated with high accuracy by quadrature divide-by-two circuits.

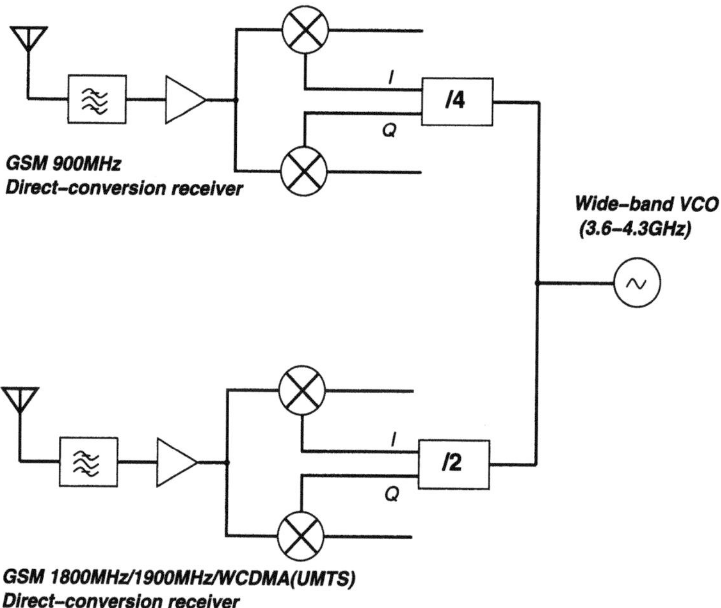

Figure 4.1. Receiver architecture for multi-band and multi-standard cellular applications.

A 2.4GHz WLAN receiver can be implemented with a direct-conversion or wide-band IF architecture as shown in Figure 4.2(a) and (b), respectively. The direct-conversion WLAN receiver use a VCO operating at 5GHz and a divide-by-2 circuit for quadrature LO signal shown in Figure 4.2(a). Alternatively, the

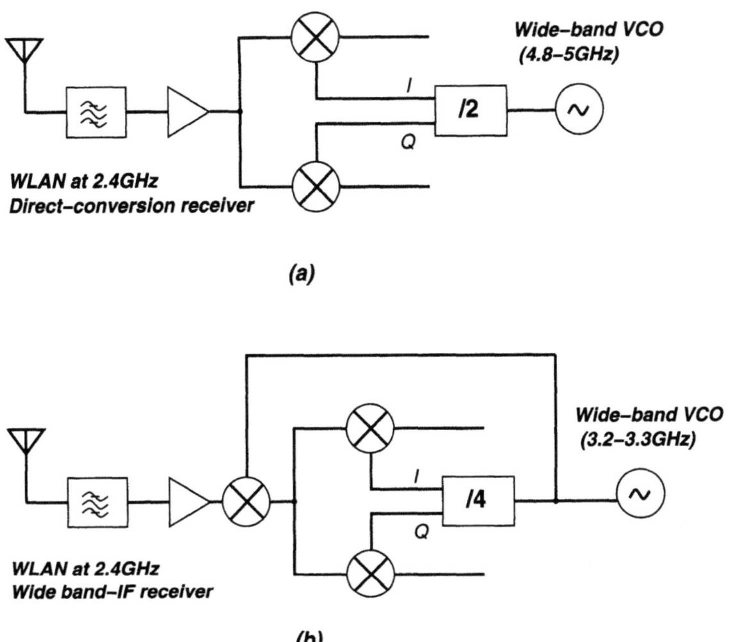

Figure 4.2. Receiver architectures for WLAN applications at the 2.4GHz ISM band.

receiver architecture in Figure 4.2(b) with two-step down conversion is using a first LO frequency at 3.2GHz which translates the 2.4GHz RF input signals to an intermediate frequency (IF) at 800MHz. Quadrature LOs at 800MHz are generated by a divide-by-four circuit from the same broadband VCO. The image frequency is located at 4GHz. Thus, in-coming RF and image signals are separated by 1.6GHz which allows for on chip filtering of the image signal [32]. The wide-band IF receiver requires the VCO to operate over the 3.2-3.3GHz range.

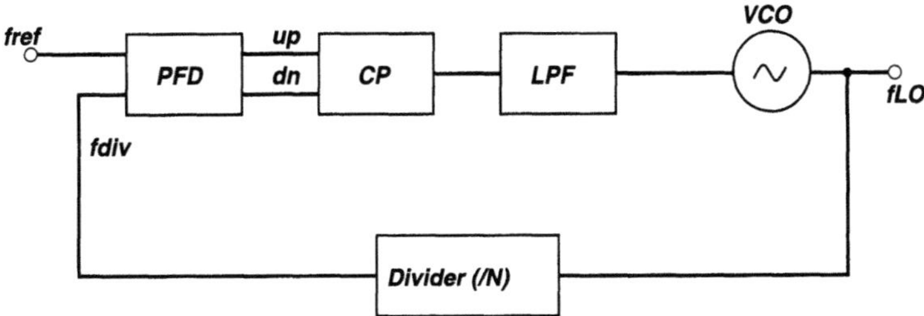

Figure 4.3. An integer-N PLL frequency synthesizer architecture.

An RF VCO is placed in a PLL for frequency synthesize as shown in Figure 4.3. The PLL consists of a VCO, a feedback divider, a phase-frequency detector (PFD), a charge pump (CP), and a low-pass filter (LPF). A low noise local oscillator (LO) signal at f_{LO} is generated by locking a noisy RF VCO to a low noise reference signal which is generated by a crystal oscillator. An integer-N PLL frequency synthesizer output frequency resolution is restricted to integer multiples of the reference frequency. This limits the loop bandwidth, and hence PLL performance, lock time, spur level, and phase noise. A more complex PLL architecture must be used to meet standard-specific characteristics such as lock time and channel spacing [33]. As an example, a hybrid integer-N/fractional-N PLL architecture is presented in more details in Chapter 8.

2.   CMOS System Integration

Monolithic RF VCOs were traditionally designed and implemented in Bipolar and GaAs technologies due to the need for RF transistors [13]. In the last decade, the scaling of the gate length in CMOS technology toward submicron values allowed implementation of RF functions. 0.25μm and further scaled CMOS technologies allow us implementation of RF transistors with peak f_T (unity current gain frequency) values excess of 30GHz. The design of monolithic VCOs in CMOS technology have been the subject of on-going research [9, 10, 34]. Fully integrated PLLs in CMOS technologies have been reported [35, 36].

The scaling of the channel length of the CMOS transistor comes with a reduction in the breakdown voltage which limits the maximum supply voltage. The decreasing supply voltage presents new challenges in the design and integration of LC VCOs at the GHz range by reducing the available tuning voltage. Integration of a RF VCO in deep sub-micron CMOS technology along with other functions requires attention in the following issues;

- The output spectral purity of VCO must be preserved in an integrated environment with a variety of noise sources from the other digital circuits and signal lines due to poor isolation and cross talk on the chip.

- Supply voltage scaling in deep sub-micron CMOS technology presents difficulties for broadband tuning range implementation due to reduced the tuning voltage.

The Spectral Purity Issue

A single chip CMOS wireless transceiver includes analog and digital functions on the same substrate. Digital circuits injects noise into the substrate when operating through switching. The noise can couple between blocks in different mechanisms; couplings through the substrate and coupling through the supply and signal lines. Noise coupled to VCO control line, supply and bias lines

will be converted into noise sidebands around the oscillation frequency through
FM, AM, and AM-FM conversion mechanisms. The AM noise contributions
are less significant for close in spectral purity of the oscillator compared to PM
or FM noise contributions. Also, they can be removed with a limiter. However, the AM noise is often converted to FM noise through AM-FM conversion.
FM and AM-FM conversion noise contributions are explored for integration
considerations. Spurious sidebands due to deterministic signals such as crystal
clock are also discussed.

FM noise contributions

The noise coupled onto the control voltage (VCNT) applied across the varactor of the LC-tank varies the tank capacitance and hence the resonant frequency.
This can be viewed as frequency-modulation (FM) noise mechanism which
translates any low-frequency noise around the oscillation frequency. The phase
noise contribution of any broadband noise source with rms value of V_n , on the
VCO control line can be described using narrow-band FM approximation [11]
by:

$$L_{VCNT}(f_m) = 10 \log \left(\frac{P_{noise}}{P_{carrier}} \right) = 10 \log \left[\frac{(K_{VCO}V_n)^2}{2f_m^2} \right] \qquad (4.1)$$

The phase noise contribution from the control line to the total PLL output
can be minimized by making the VCO gain, K_{VCO}, as small as possible. The
noise sources on the control line are loop filter resistor thermal noise and CP
current noise in addition to any other noise coupled from the substrate and signal
routings . The loop filter resistor's thermal noise contribution to the PLL output
phase noise can be computed by using the models developed in Chapter 3.

In general, the FM noise mechanism is applicable to any noise source which
can vary (or modulate) the frequency of the oscillation [11]. The VCO output
frequency is sensitive to bias current and supply voltage variation. Any noise
associated with the bias current or the supply voltage will also contribute to the
VCO phase noise through FM noise mechanism in a similar fashion. Leeson's
Equation 3.14 can be extended to include these noise contributions [37];

$$\begin{aligned} L(f_m) \;=\; & 10 \log \left[\frac{2FkT}{P_s} \left(\frac{f_o}{2Q_L f_m} \right)^2 \left(1 + \frac{f_k}{f_m} \right) \right. \\ & \left. + \; \frac{|K_{VCO}|^2}{2f_m^2} S_{VCNT} + \frac{|K_{VDD}|^2}{2f_m^2} S_{VDD} + \frac{|K_{IB}|^2}{2f_m^2} S_{IB} \right] \end{aligned} \qquad (4.2)$$

where S_{VCNT} and S_{VDD} are the voltage noise spectral densities of the control
voltage and supply voltage in V^2/Hz, respectively. K_{VCO} and K_{VDD} are
defined as the sensitivity of the frequency of oscillation for control voltage

and supply voltage [37], respectively. K_{IB} is the sensitivity of the oscillation frequency to the bias current variations [38–40].

Therefore, the bias and supply sensitivity of the VCO should be minimized for low noise. The bias and supply noise can be minimized by filtering without affecting the PLL dynamics which can not be done for the noise associated with the tuning control input of the VCO. The layout and circuit techniques play important roles in reducing the noise coupled to the tuning line on the chip.

AM-FM conversion noise contribution

The effective capacitance of a varactor depends on the DC voltage and amplitude of the oscillation [41] [42] [43]. The various noises coupled from supply, bias, and substrate modulate the amplitude of the oscillation. As an example, a differential LC oscillator up-converts low-frequency bias noise into AM sidebands around the oscillation frequency [42]. Usually, the AM noise represents smaller contribution at close offset frequencies compared to FM (or PM) noise contributions. Also, it can be removed with a limiter circuit [42]. On the other hand, the AM noise coupled to resonator also modulate the effective capacitance of a varactor which then converts the AM noise into FM noise. The FM noise sidebands are indistinguishable from the phase noise, and a limiter cannot suppress FM or phase noise [42]. The AM-FM noise contribution can be calculated using narrow-band FM approximation [42];

$$L_{AM-FM}(f_m) = 10 \log \left[\frac{(K_{AM-FM} V_{n,AM})^2}{2f_m^2} \right] \tag{4.3}$$

where K_{AM-FM} is the AM-FM conversion gain in V/Hz. $V_{n,AM}$ is the rms AM noise voltage spectral density in $V/\sqrt{Hz}$. The AM-FM conversion gain is defined as [42];

$$K_{AM-FM} = \frac{\partial \omega_o}{\partial A} \tag{4.4}$$

where A is the amplitude. Since $\omega_o = 1/\sqrt{LC_{eff}}$, then

$$K_{AM-FM} = -\frac{1}{2} \frac{\omega_o}{C_{eff}} \frac{\partial C_{eff}}{\partial A} \tag{4.5}$$

The sensitivity of a MOS varactor's C_{eff} to oscillation amplitude is given by [42],

$$\frac{\partial C_{eff}}{\partial A} = \frac{C_{max} - C_{min}}{\pi} \frac{2V_{DC}}{A^2} \sqrt{1 - \left(\frac{V_{DC}}{A} \right)^2} \tag{4.6}$$

From Equations 4.5 and 4.6,

$$K_{AM-FM} \propto C_{max} - C_{min} \tag{4.7}$$

Equation 4.7 suggests that a wide tuning range results in a larger AM-FM gain, and in-turn larger AM-FM noise conversion. A similar relationship for a p^+/n junction diode can be derived [41].

Spurious Sidebands

Also, any deterministic signal with frequency, f_s, on the control line (such as a reference signal feeding through from the CP output and the crystal clock coupled to the tuning line), will be up-converted as discrete sidebands around the oscillation frequencies; $f_o \pm f_s$. These side bands are called *reference spurs* which is due to the reference signal coupled the tuning line. The magnitude of the spurious side bands with respect to the carrier magnitude can be calculated using Equation 4.1 by replacing V_n by $V_s/\sqrt{2}$ for a periodic signal, $V_s \cos \omega_s t$.

$$L_{spur}(f_m) = 10 \log \left(\frac{P_{spur}}{P_{carrier}} \right) = 10 \log \left[\frac{(K_{VCO} V_s)^2}{4 f_s^2} \right] \quad (4.8)$$

The spurious sidebands of VCO can be greatly reduced by minimizing VCO the gain. The spurious sidebands due to an on-chip crystal clock can also be reduced by using the largest possible crystal frequency, f_{XCO}. The PLL reference frequency can be generated by using divider from the crystal clock signal.

Another way to minimize spurious sidebands is to better isolation of the tuning line. The integrated loop filter placed on chip close the VCO can also improve the isolation of the tuning line from the unwanted disturbances. The on-chip loop filter implementation is not always possible when the components values are large for physical implementation on the chip. The off-chip placement of the loop filter usually results in long routing lines to the VCO, and it requires more attention at the top layout.

Supply Voltage Issue

The VCO tuning range must cover all selectable channels within a desired charge-pump (CP) output voltage (or VCO control voltage, VCNT) in the presence of the PVT variation. A typical VCO and CP connection through loop filter is shown in Figure 4.4(a). Desired tuning curve is shown under the shaded area of Figure 4.4(b). Implementing such a function shown in Figure 4.4(b), however, presents challenges in sub-micron CMOS technology due to scaled supply voltage which limits the available tuning voltage. It is still possible to implement a wide tuning range with a small tuning voltage range using MOS devices as varactors [44] [45] [46]. However, this will result in a large VCO gain which makes the VCO very sensitive to noise and disturbances on the tuning

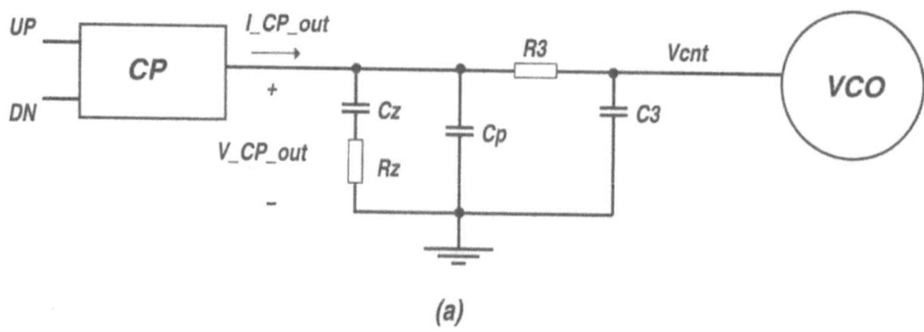

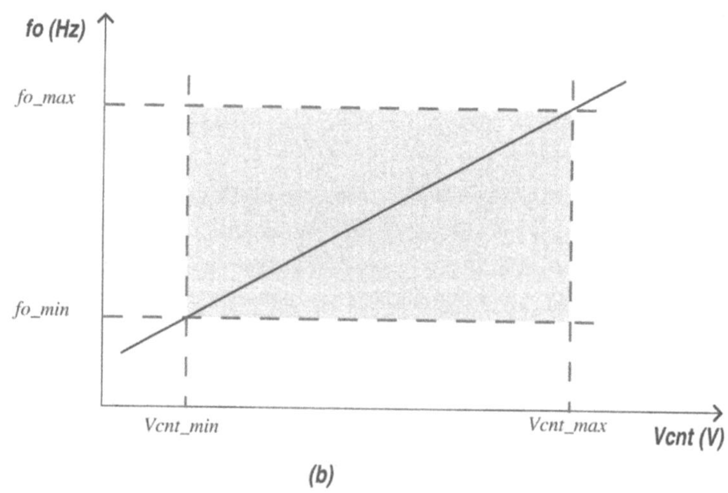

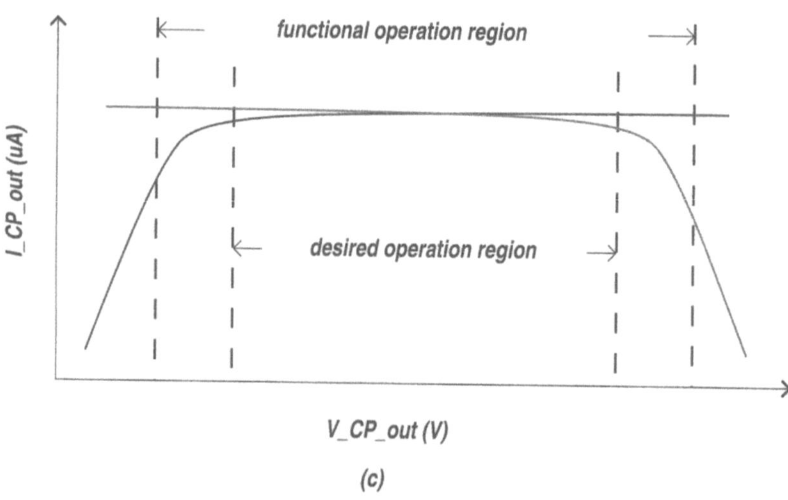

Figure 4.4. (a) VCO and CP connections through loop filter (b) Desired operation region for VCO tuning curve (c) CP output current as a function of output voltage.

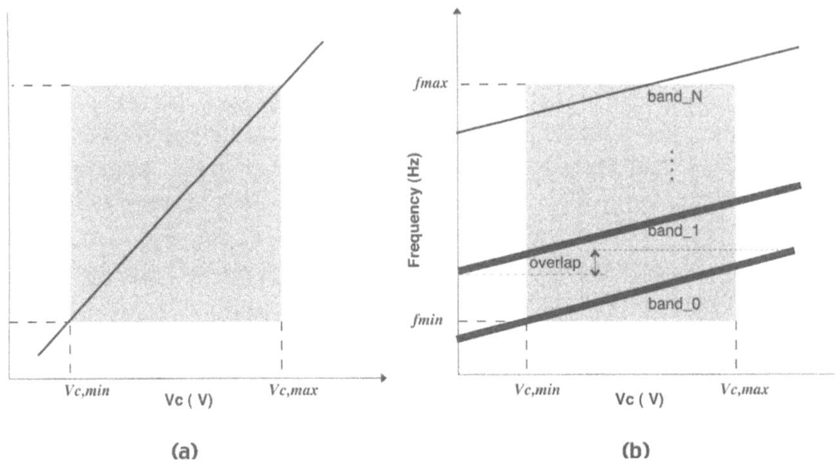

Figure 4.5.   (a) single continiuos tuning curve (b) tuning curve divided into sub-bands.

line as explained in Section 2.0. This is especially important for fully integrated transceiver solutions where blocks are poorly isolated due to conductive substrate.

As the feature size of the CMOS technology shrinks, the nominal supply voltage also shrinks. A typical $0.18\mu m$ CMOS supply voltage is 1.8 V. A typical charge pump will properly operate at an output voltage range between 0.4V-1.4V while keeping the output transistors in saturation as shown in Figure 4.4(c). For the example shown in Figure 4.1, the desired tuning range is about 700MHz. To account for process, supply voltage and temperature variation, the tuning range should be at least twice the required frequency range to be synthesized by the PLL (Figure 4.1). This translates into a 1400 MHZ/V of a VCO gain, K_V, which is large and will cause an increase of the noise at the PLL output.

In addition to the required minimum tuning range, the on-chip varactor C-V (capacitance-versus-voltage) characteristics do not exhibit a linear behavior. This results in a non-linear VCO tuning curve and the K_V varies with the control voltage. The tuning curve nonlinearity, i.e. varying VCO gain over the tuning voltage, can degrade the overall PLL performance parameters.

One approach to achieve a wide tuning range with a small VCO gain and linear tuning curve simultaneously is to break a single broadband tuning curve into several narrow sub-bands with sufficient frequency overlap. This can be implemented by using both discrete and continuous tuning as shown in Figure 4.5(b). The continuous tuning can be implemented by using a reverse-biased p^+/n junction or MOS varactor. One method of discrete tuning scheme has been exploited previously in [34]. Circuit techniques to achieve discrete tuning scheme for tuning range with sub-bands are explored in the next chapter.

3.   Summary

In this chapter, the system design considerations for broadband CMOS VCO design are discussed at the system level. The radio and frequency synthesizer architectures must be designed concurrently around the frequency plan for multi-band and multi-standard applications. The design parameters for broadband VCOs are phase noise, tuning range and power consumption as an individual block. CMOS implementation of the broadband VCO also requires attention for on-chip noise isolation and robust design against process variation for single chip radio solutions.

Chapter 5

BROADBAND VCOs:
CIRCUIT DESIGN CONSIDERATIONS

This chapter deals with circuit design considerations for broadband RF CMOS VCOs as well as VCO implementation in fully integrated transceiver solutions. RF VCO with subbands is considered for broadband implementation. Discrete and continuous tuning schemes for broadband VCOs are explored in details. Capacitance and inductance switching techniques for discrete tuning control are evaluated for performance and implementation considerations.

Both the active circuit and resonator tank design for RF VCO design are overviewed. Bias circuit is a major noise contributor to the VCO phase noise. Bias filtering techniques are investigated. Dynamic bias filtering solutions are presented for fast VCO start-up. Passive components forming the LC tank of the VCO are of paramount importance since they mainly determine noise performance and power consumption of the VCO. Integrated passives, inductors and varactors, are explored for high performance solutions.

1.   Broadband VCO with Subbands

As pointed out, it is possible to design a VCO with a wide tuning range using single tuning band as shown in Figure 4.5(a) by employing MOS devices as varactors [44] [45] [46]. This implementation, however, is not suited for monolithic integration due to high VCO sensitivity and tuning nonlinearity. Broadband VCO implementation with subbands as shown in Figure 4.5(b) is considered here for integration purposes for lower VCO sensitivity and improved tuning linearity. Broadband VCO operation with subbands can be implemented with one of following approaches;

1 Switching in or out discrete amounts of capacitance or inductance from the LC tank as shown in Figure 5.1(a). C_{VC} in Figure 5.1(a) is capacitance of

the varactor used for continuous tuning of the VCO frequency through the control or tune signal from the PLL.

2 Switching between LC tanks which are separately optimized and tuned for different frequency bands as shown in Figure 5.1(b). This method is often used in discrete broadband RF VCO design [47]. This method has advantages over the first method since each LC-tank can be optimized for low phase noise in its relevant operating band. However, this method is not suited for integrated solutions mainly due to unavailability of high quality integrated RF switches and the large die often taken by integrated inductors.

3 Switching between VCOs which are optimized for different frequency bands as shown in Figure 5.1(c). The VCO of a desired band is enabled while the other VCOs are disabled. The buffers provide isolation for the VCOs. This is not well suited for integrated solutions because of large the die area taken by several VCOs.

Switching capacitance or inductor inside the LC tank is preferred solution for fully integrated VCO because it requires only one LC-tank with digitally controllable components. In general, if the frequency band switching required isn't very large (within 30%), it may usually be realized within the same tank circuit, by switching on or off an additional capacitor or inductor. However, if the required switching is more than 30%, it becomes very difficult to satisfy both broadband and low noise requirements in a single design. Switching between LC tanks or VCOs is necessary for low noise performance when the tuning range is 30% or more. Alternatively, if the tuning range consists of discrete bands which are separated by a 50% difference, then a divide-by-two circuit and a single band VCO can be used.

2. Switching Techniques for Broadband Operation

Switching a discrete amount of inductance or capacitance from a LC-tank to design broadband VCO is overviewed. The design trade-offs are investigated.

Capacitance Switching

Switching a capacitance to extend tuning range is first proposed by Kral [34]. In CMOS technology, n-type MOS transistors are used to build an RF switch. Equivalent circuits of an n-type MOS transistor to switch in and out a capacitance value of C_o to the resonator tank at high frequencies are shown in ON and OFF states in Figure 5.2(a),(b), respectively. C_d is the total drain parasitic fringe capacitance which is equal to $W_{sw}C_{dd}$. W_{sw} is the width of the switching transistor and C_{dd} is the drain fringe capacitance with a unit of $fF/\mu m$.

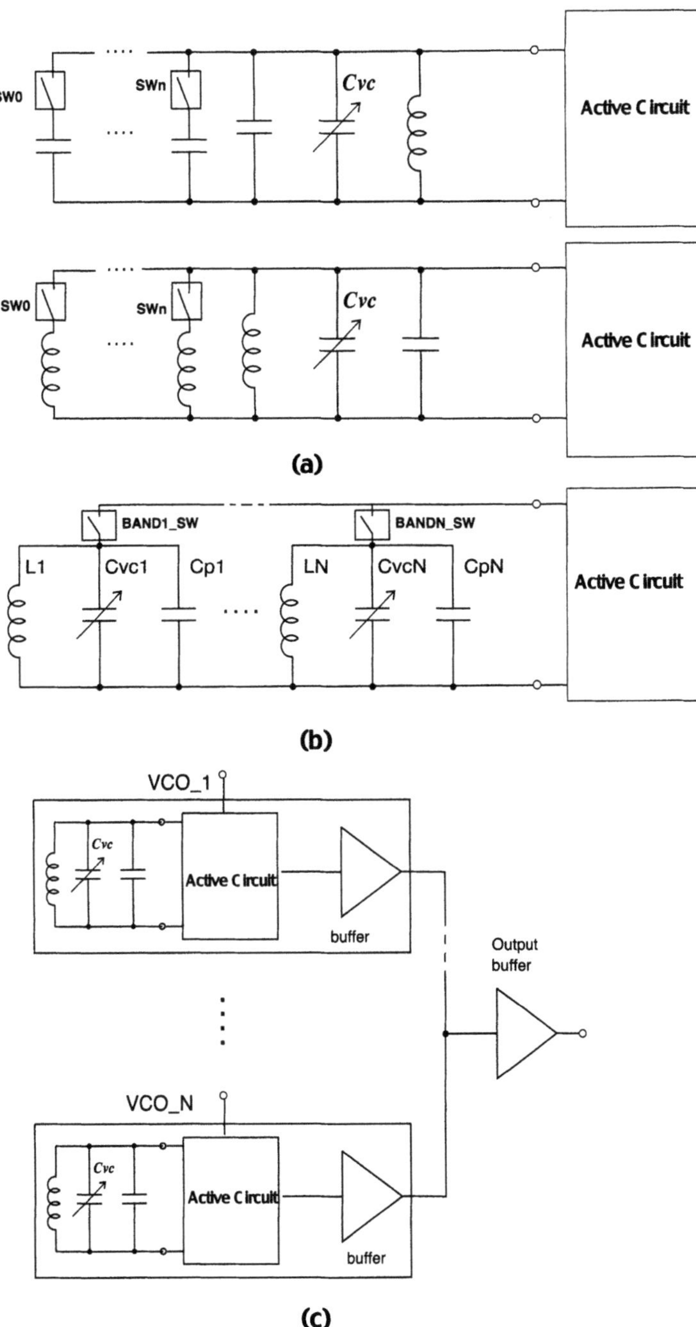

Figure 5.1. Different approaches for broadband VCO with subbands (a) switching inside the LC-tank (b) switching between LC-tanks (c) switching between VCOs.

In the OFF state, the equivalent circuit for the switch and the capacitance (C_o) are shown in Figure 5.2(a). The impedance of the circuit in the OFF state assuming $R_{OFF} >> (\omega C_d)^{-1}$ and $C_{OFF} = C_o//C_d$ is written as,

$$Z(s) = \frac{1}{sC_{OFF}} \tag{5.1}$$

In the ON state, the equivalent circuit for the switch and the capacitance (C_o) are shown in Figure 5.2(b). The channel resistance, R_{ON}, of the switching transistor is added in series to the capacitance C_o. The impedance of the circuit in the ON state assuming $R_{ON} << (\omega C_d)^{-1}$ is written as,

$$Z(s) = R_{ON} + \frac{1}{sC_o} \tag{5.2}$$

The resistance R_{ON} is the channel resistance of the MOS transistor, and is written as,

$$R_{ON} = \left[(\mu_n C_{ox})(\frac{W}{L})(V_{GS} - V_t) \right]^{-1} \tag{5.3}$$

As CMOS technology scales down, R_{ON} will improve since the channel length is inversely proportional to the channel resistance.

Two performance parameters are of interest for design consideration; (i) the quality factor of the tuning circuit at the frequency of operation (ii) the ratio of the maximum to the minimum capacitance. The quality factor of the switched capacitance is lowest when the switch device is ON. The quality factor for the ON state is written as,

$$Q = \frac{1}{\omega_o C_o R_{ON}} \tag{5.4}$$

where R_{ON} is given by 5.3 and ω_o is the operating frequency. From Equations 5.3 and 5.4, the maximum quality factor dependency is given as,

$$Q \propto \frac{W_{sw}}{\omega_o C_o} \tag{5.5}$$

where W_{sw} is the width of the switching transistor. For maximum quality factor, $L = L_{min}$ and $V_{GS} - V_t = V_{DD} - V_t$ are chosen according to Equation 5.3. The parameters, W_{sw} and C_o, are design variables.

The tuning range is dependent on the ratio of the maximum to the minimum capacitances since the oscillation frequency is proportional to $1/\sqrt{LC}$. This ratio is written as;

$$\frac{C_{max}}{C_{min}} = \frac{C_o}{W_{sw} C_{dd}} + 1 \tag{5.6}$$

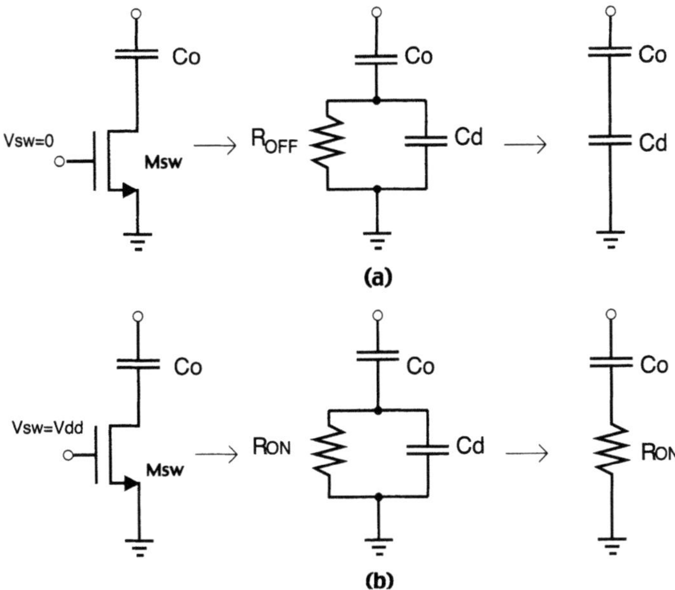

Figure 5.2. NMOS transistor as an RF switch (a) VSW is low, the OFF state (b)VSW is high,the ON state.

Equations 5.5 and 5.6 suggest that maximizing the quality factor of the switched capacitance conflicts with maximizing the tuning ratio. The width of the switching transistor, W_{sw}, is the key design parameter for optimization when the switched capacitance amount, C_o, is given with the operating frequency, ω_o.

The quality factor of the switched capacitance can be improved by by using a single switch device for two capacitances in a differential fashion [9] as shown in Figure 5.3. One half of the resistance, R_{ON}, is added to each capacitance when the switch is ON. This way the quality factor, Q, would improve twice as much compared to the single ended structure of Figure 5.2.

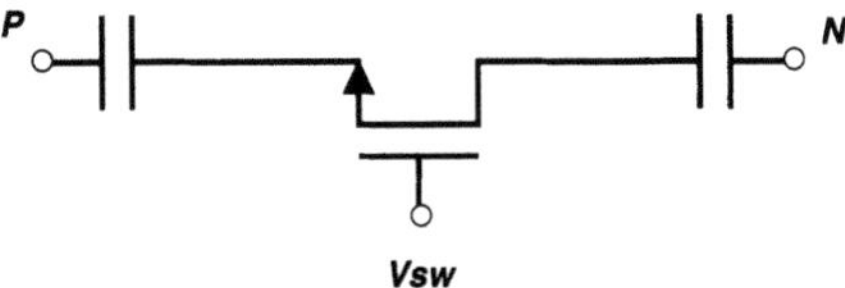

Figure 5.3. Differential capacitance switching.

Two different bias configurations for differentially switched capacitances are shown in Figure 5.4(a) and(b). The only difference between them is the biasing of the MOS switch device, M_1. The resistors R_1 and R_2 in Figure 5.4 (a) are used for DC biasing of the S/D terminals of the MOS switch device. The value of these resistors should be chosen large enough for the operation frequency to exhibit high impedance for RF signals. When V_{sw} is set to 0, $V_{GS} - V_t$ of the MOS switch device is maximized by setting $V_{D/S}$=0V and $V_G = V_{DD}$. The equivalent capacitance seen from differential ports takes maximum value leading to a lower frequency operation. When V_{sw} is set to V_{DD}, the upper frequency band is chosen. $V_{D/S} = V_{DD}$ and $V_G = 0V$ are voltage values for higher voltage setting.

The resistors R_1 and R_2 in Figure 5.4(a) are replaced by minimum-size MOS devices, M_2 and M_3 which act as active resistor operating at the sub-threshold region (Figure 5.4(b)). The active MOS resistors occupy smaller die area, and hence less parasitic capacitances are added to critical RF nodes.

The switched capacitance is usually implemented by using MIM (metal-insulator-metal) capacitance, and is called the MIM cap switching.

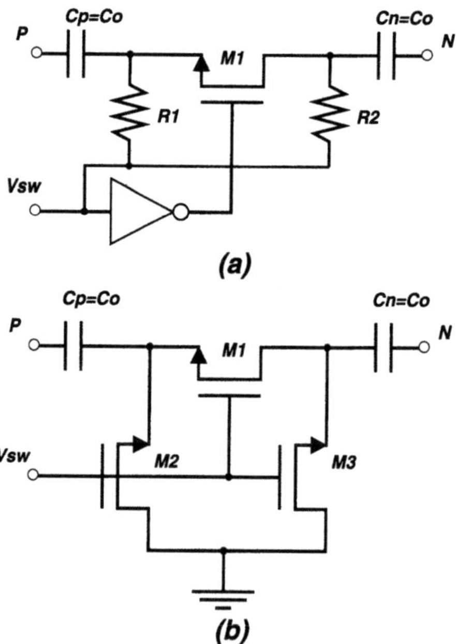

Figure 5.4.   Differential capacitance switching (a) with resistor biasing (b) with MOS biasing.

Design of MIM capacitance switching circuit

The total LC-tank capacitance is formed by fixed parasitic capacitances (C_{par}, from active devices, inductor, and interconnects), continuous tuning varactor (varies from $C_{v,min}$ to $C_{v,max}$), and a switching MIM cap array (varying from $C_{sw,min}$ to $C_{sw,max}$). A three-bit switching circuit with continuous tuning varactor is shown in Figure 5.5(a). The switch MOS devices' widths and the MIM cap values are binary weighted, i.e., $W_o, 2W_o, \ldots, 2^{N-1}W_o$ and $C_o, 2C_o, \ldots, 2^{N-1}C_o$ to obtain equal switching step size. The minimum switching capacitance value is calculated as,

$$C_{sw,min} = \left(\frac{1}{C_o} + \frac{1}{C_{d,o}}\right)^{-1} + \left(\frac{1}{2C_o} + \frac{1}{2C_{d,o}}\right)^{-1} + \cdots \left(\frac{1}{2^{N-1}C_o} + \frac{1}{2^{N-1}C_{d,o}}\right)^{-1} \tag{5.7}$$

or,

$$C_{sw,min} = (2^N - 1)\left(\frac{1}{C_o} + \frac{1}{C_{d,o}}\right)^{-1} \tag{5.8}$$

and the maximum switching capacitance value is calculated as,

$$C_{sw,max} = (2^N - 1)C_o \tag{5.9}$$

where C_o and $C_{d,o}$ are the unit MIM capacitance value and the drain fringe capacitances value of the unit sized device (W_o, L_o), respectively. The frequency overlap inequality can be written as,

$$C_{v,max} - C_{v,min} > C_o - \left(\frac{1}{C_o} + \frac{1}{C_{d,o}}\right)^{-1} \tag{5.10}$$

The maximum and minimum capacitances in the resonator can be calculated as,

$$C_{res,max} = C_{v,max} + (2^N - 1)C_o + C_{par} \tag{5.11}$$

$$C_{res,min} = C_{v,min} + (2^N - 1)\left(\frac{1}{C_o} + \frac{1}{C_{d,o}}\right)^{-1} + C_{par} \tag{5.12}$$

Capacitance Switching with MOS devices

MOS devices exhibit a very steep C-V profile due to the fast transition from depletion (Cmin) to strong inversion (or accumulation) (Cmax) [45, 48]. MOS devices can be used for discrete capacitance switching by using the gate as one terminal and the drain/source tied together as the other terminal. Changing the voltage digitally from 0 to V_{DD} between the two ports of the device, the capacitance seen across the terminals of the device can be changed from depletion to inversion (or accumulation).

The ratio of the maximum to the minimum capacitance heavily depends on the technology. The maximum capacitance occurs in the accumulation (or inversion) case, and it is determined by the gate-oxide thickness. The minimum

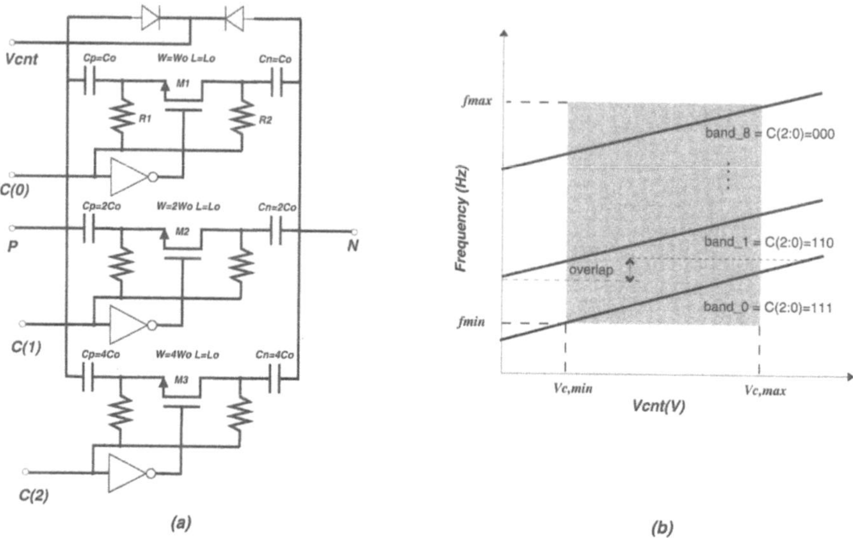

Figure 5.5.   Band switching circuit (a) simplified schematic (b) Tuning curves.

capacitance occurs in the depletion case. The capacitance can be calculated for these cases by using the following equations.

$$C_{ox} = \frac{\varepsilon_{SiO_2}\varepsilon_o}{t_{ox}} \tag{5.13}$$

$$C_{dep} = \frac{\varepsilon_{Si}\varepsilon_o}{d} \tag{5.14}$$

where ε_o is the permittivity of free space, t_{ox} is the gate oxide thickness, ε_{SiO_2} is the dielectric constant of gate oxide, ε_{Si} is the dielectric constant of silicon (n-well), and d is the maximum depletion width which is dependent on the n-well impurity concentration for a specific process. The maximum to minimum capacitance ratio improves as CMOS technology scales down since the gate oxide becomes thinner. Equations 5.13 and 5.14 give an insight on the capacitance values. The actual capacitance can be obtained from SPICE simulation for a given bias condition. The channel resistance of a PMOS device in strong inversion is written as [49, 50],

$$R_{ch,PMOS} = \left[(12\mu_p C_{ox})(\frac{W}{L})(V_{GS} - |V_{tp}|)\right]^{-1} \tag{5.15}$$

The drawbacks of a MOS capacitance are the nonlinearity of the C-V curve and modeling at high frequencies. The nonlinearity also causes the low frequency noise from the control line to be converted to phase noise through FM mechanism. Nonetheless, the MOS capacitance switching offers some advantages over MIM capacitance switching:

- Since the gate-oxide is typically the thinnest layer in the process, it has higher density than the MIM capacitance. The gate oxide capacitance density is about $8fF/\mu^2$ while the MIM capacitance density is about $1fF/\mu^2$ in 0.18μ CMOS technology. Hence, a MOS switch will result in much smaller die area and smaller parasitic capacitance loading the tank than a MIM cap switching.

- The gate-oxide thickness is better controlled than the oxide thickness of a MIM capacitance. The gate-oxide capacitance variation is in the order of $\pm5\%$, while the MIM capacitance variation is in the order of $\pm20\%$.

- The MOS is available in a standard CMOS process while the MIM cap is optional and may not be available in every technology. Hence, MIM capacitance use may increase cost.

- The maximum Q of a MIM cap switching and MOS cap switching are limited by the switching device ON resistance (Equation 5.3) and the channel resistance (Equation 5.15). The MOS capacitance can be used at higher frequencies than the MIM cap switching in the same technology since the Q of the MOS capacitance is proportional to $1/(\omega R_s C_s)$.

Design of MOS capacitance switching circuit

The total LC-tank capacitance is formed by fixed parasitic capacitances (C_{par}, from active devices, inductor, and interconnects), continuous tuning varactor (varies from $C_{v,min}$ to $C_{v,max}$), and a switching MOS cap array (varying from $C_{sw,min}$ to $C_{sw,max}$). A three-bit switching circuit with continuous tuning varactor is shown in Figure 5.6(a). The switched MOS devices' widths are binary weighted, i.e., $W_o, 2W_o, \ldots, 2^{N-1}W_o$ to obtain equal switching step size with N being number of units. The minimum and maximum switching capacitance for array of N devices can be calculated from;

$$C_{sw,min} = (2^N - 1)\left(\frac{1}{C_{ox,o}} + \frac{1}{C_{dep,o}}\right)^{-1} \tag{5.16}$$

$$C_{sw,max} = (2^N - 1)C_{ox,o} \tag{5.17}$$

where $C_{ox,o}$ and $C_{dep,o}$ are the oxide capacitance and depletion capacitances of the unit sized device (W_o, L_o), respectively. The frequency overlap inequality can be written as,

$$C_{v,max} - C_{v,min} > C_{ox,o} - \left(\frac{1}{C_{ox,o}} + \frac{1}{C_{dep,o}}\right)^{-1} \tag{5.18}$$

The maximum and minimum capacitances in the resonator can be calculated as,

$$C_{res,max} = C_{v,max} + (2^N - 1)C_{ox,o} + C_{par} \qquad (5.19)$$

$$C_{res,min} = C_{v,min} + (2^N - 1)\left(\frac{1}{C_{ox,o}} + \frac{1}{C_{dep,o}}\right)^{-1} + C_{par} \qquad (5.20)$$

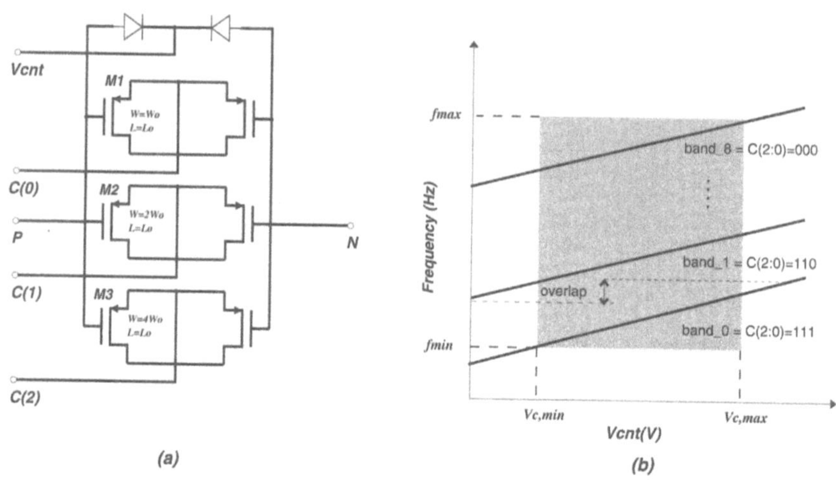

Figure 5.6.   Band switching circuit (a) simplified schematic (b) Tuning curves.

Inductance Switching

Switching inductors in the LC-tank are also feasible and shown in [51, 52]. Single ended and differential switching is shown in Figure 5.7(a) and (b), respectively. Differential switching of inductor has the benefit of reduction of series resistance and parasitic capacitance, by a factor of 2, to the RF nodes in the tank.

Switching the inductors suffers from the CMOS switch loss as in the case switching capacitance. The switch loss is more pronounced in the inductor case since the quality factor of an on-chip inductor is already smaller than that of the capacitance. If the switches are designed for minimum series resistance to avoid deteriorating of the Q of the tank, then a large parasitic switch capacitance is introduced at the RF-nodes of the tank which leads to high power consumption and phase noise.

Implementation of equally spaced subbands as shown in Figure 4.5(b) is difficult with switching the inductors. This requires an implementation of a binary-weighted inductor switching structure which in turn implies implementing several inductors, L_o, $2L_o$, $4L_o$,... Characterization of such a structure is

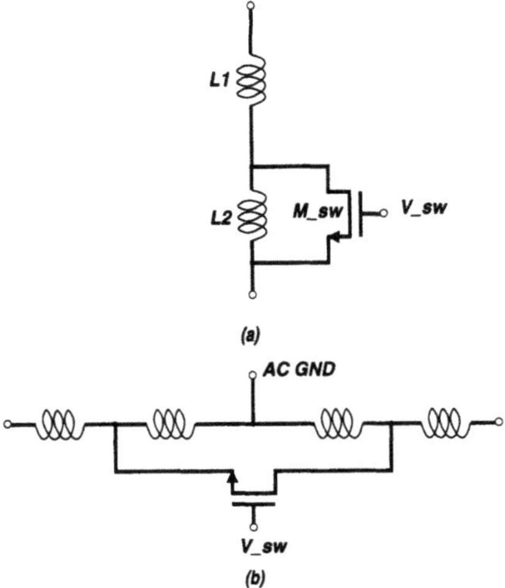

Figure 5.7. Inductor switching (a) Single ended inductor switching reported in [51] (b) Differential inductor switching reported in [52].

a prohibiting task. Furthermore, it will occupy a large die area. Therefore, inductor switching does not allow digitally controllable subbands.

The tuning curves linearity is not well controlled in inductor switching with large steps. The oscillation frequency of a VCO is proportional to $(L[C_p + C(v)])^{-1/2}$. The variable capacitance, $C(v)$, in the resonator stays same when switching from a small value to a large inductor value.

The inductor switching implementation in [51, 52] demonstrates switching between 900MHz and 1800MHz bands. For all, practical purposes, the 900MHz band frequencies can be obtained from the 1800MHz band with a divide-by-two circuit.

3. Broadband VCO Implementation

A typical CMOS LC-VCO employs a differential topology with cross-coupled NMOS, PMOS, or complementary NMOS and PMOS pairs to realize negative resistance to compensate the loss of the resonator tank. The differential topology has advantages in a fully integrated circuit realization, such as rejection of common-mode supply and substrate noise, and differential outputs. The negative resistance is created by the positive feedback of the cross coupled transistors.

Active Circuit Design

The versions of the differential topology with cross-coupled pair are shown in Figure 5.8. The selection of a topology depends on specifications of the VCO (phase noise, power consumption, and tuning range) and the process technology.

The NMOS only topology shown in Figure 5.8(a) has the advantage of higher transconductance per area compared to PMOS only topology shown in Figure 5.8(b) due to higher mobility of electrons in NMOS devices, and hence smaller transistor capacitances will contribute to the total parasitic capacitance of the resonator tank. However, PMOS devices have lower flicker noise densities than NMOS devices.

For the NMOS and PMOS complementary topology (Figure 5.8(c)), power consumption is lower than that of the NMOS only or PMOS only topologies because the bias current is reused. This topology is use-able where there is enough supply voltage headroom for stacking the transistors. As the channel length shrinks with sub-micron CMOS technology, the available supply voltage does shrink but not V_t of MOS devices. The nominal supply voltages for 0.18um and 0.13um CMOS technologies are 1.8V and 1.25V, respectively. The nominal threshold voltages for these technologies are 0.5V and 0.4V, respectively.

The topologies shown in Figure 5.8 use current biasing for active devices. The current biasing serves two purposes; (i) limiting the output amplitude of VCO, and hence preventing devices from going into deep triode region which degrades phase noise performance [53] (ii) presents high impedance to the node connected to the resonator to decouple supply or ground from the resonator. The bias current can be suplied from the supply or ground sides. If there is available headroom, bias current can be enforced from both sides. Usually, the supply side is preferred to reduce the supply sensitivity of the output frequency of the VCO. Unfortunately, the bias current noise is one of the major contributors to the phase noise of the VCO. The VCO acts as mixer for bias noise and translates low frequency bias noise into sidebands around oscillation frequency.

The small signal admittance seen at the the drain of M1 and M2 transistors in Figure 5.8(a) and (b) pairs is equal to $G_{in} = -g_m/2$, where g_m is the small-signal transconductance of each transistor. The MOS transconductance, g_m, is written in the saturation region as;

$$g_m = (\mu_n C_{ox})(\frac{W}{L})(V_{GS} - V_t) \tag{5.21}$$

where $(\mu_n C_{ox})$ is a fixed process dependent design parameter. The transistor geometry, $(\frac{W}{L})$, and the DC bias point, $(V_{GS} - V_t)$, are chosen to achieve a desired g_m.

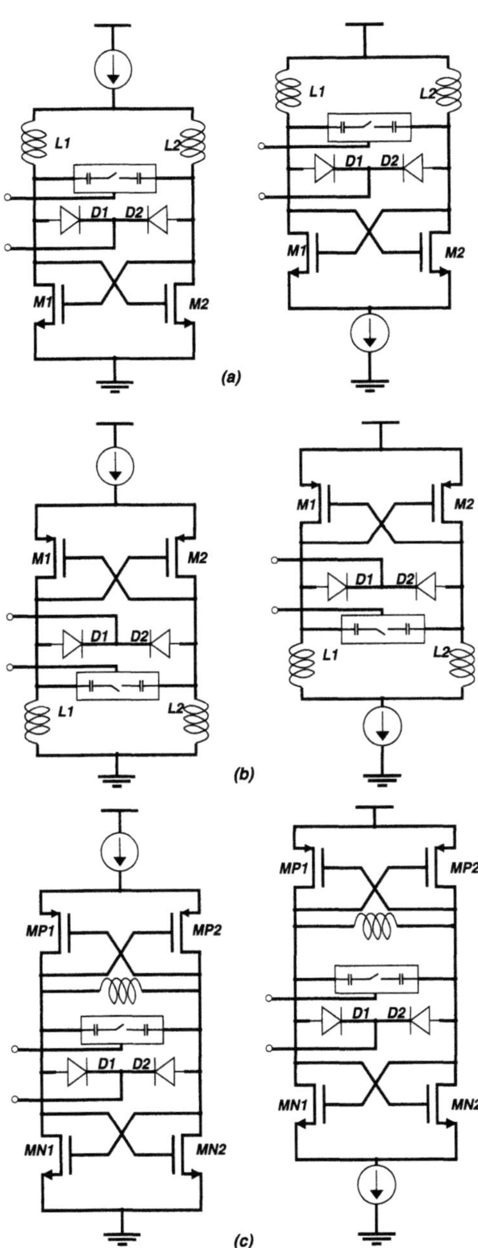

Figure 5.8. VCO topologies (a) NMOS only (b) PMOS only (c) NMOS and PMOS complementary.

A trade-off exists between power savings, tuning range, and phase noise performance when choosing transistor size, $(\frac{W}{L})$ and bias point $(V_{GS} - V_t)$. To lower power consumption, $(V_{GS} - V_t)$ has to be as low as possible. In that case, $(\frac{W}{L})$ must be increased to obtain a desired g_m value. This will lead to large parasitic capacitances adding to the capacitance of the resonant tank which results in reducing the available tuning range.

Furthermore, minimizing $(V_{GS} - V_t)$ and $(\frac{W}{L})$ value to reduce power consumption and increase tuning range will yield a small oscillation amplitude $(V_{osc} \propto I_{osc})$. The smaller oscillation amplitude results in poor phase noise performance since phase noise is inversely proportional to the square of oscillation amplitude, i.e., $PN \propto 1/V_{oscamp}^2$.

Minimizing the power consumption in aVCO requires sacrifices from tuning range and phase noise peformance. Maximizing tuning range yields higher power consumption and phase noise. Minimum phase noise requires narrow tuning range (larger L/C ratio) and highest output amplitude (power consumption).

The negative resistance seen at the active circuit port (Figure 5.8), $R_a = -2/g_m$, must be chosen at least two times greater than the resonant tank effective resistance to guarantee the start up of oscillation [11, 13].

Programmable Bias Current

A VCO bias current can be made programmable in order to dynamically change the VCO core current for VCO amplitude adjustment when the resonator Q is low due to process variation, temperature, or supply voltage variation. Also, the resonator Q varies over the tuning range for broadband VCOs. In steady-state operation, oscillator active devices operate as switches rather than a small signal negative resistance. The current fed to the resonator tank is limited by the current mirrors in the oscillator bias circuit and becomes a square wave with a peak current I_{osc}. The output amplitude of the oscillator can be approximated by using the first Fourier component in the current [54];

$$V_p = \frac{2}{\pi} I_{osc} R_p \qquad (5.22)$$

where I_{osc} is the current fed to the resonator by active devices and V_p is the peak voltage over the resonator equivalent tank resistance, R_p, at the steady-state which equals to the peak to peak value of one single-ended output. From Equation 5.22, the oscillator output amplitude is primarily dependent on the resonator equivalent resistance and the active device bias current. A simple method to keep the VCO amplitude constant or at the desired level over the frequency band of interest is to keep the product of $I_{osc} \times R_p$ constant. This can be done or implemented several ways. Automatic-amplitude control (AAC) circuit can be implemented [54, 55]. An extensive discussions of digital and analog AAC for VCOs can be found in Reference [54].

Amplitude control circuit is needed in broadband VCOs to guarantee a proper operation over a wide tuning range. Two different architectures to correct the amplitude of the broadband VCO are shown in Figure 5.9. Figure 5.9(a) shows digital amplitude correction circuit with an ADC circuit. Figure 5.9(b) shows digital current control depending on sub-band selection code in a band switching tuning scheme. An implementation of the programmable bias block is also shown in Figure 5.9(c). The TR(0:N) is the sub-band selection code in a band switching tuning scheme.

Bias Noise Filtering

Bias filtering is necessary to suppress the noise from band-gap reference current noise, noise coupled reference current line and noise from the digital control inputs in the bias circuit shown in Figure 5.9. A typical low-pass bias filter is shown in Figure 5.10(a). The $R_b \times C_b$ product determines the noise bandwidth. The noise bandwidth of the low pass filter should be less than the PLL loop bandwidth since the PLL suppresses the VCO noise inside the PLL loop bandwidth.

Typical resistors (high-resistivity poly) and capacitors (MIM and poly-poly) in CMOS technology consume large die area. A solution is proposed for this as shown in Figure 5.10(b). The resistor is implemented by using a MOS device in ohmic region while a MOS device is used as capacitor. Implementing a large resistor with MOS device takes less die area than MOS capacitor. Hence, it is desirable to make the resistance large and the capacitance small while keeping the RC product constant. The MOS device resistance in the triode region is given by;

$$R_{DS} = \left[\mu_n C_{ox} \left(\frac{W}{L} \right) (V_{GS} - V_T) \right]^{-1} \tag{5.23}$$

The resistance of the MOS device is function of $\left(\frac{W}{L} \right)$ and $(V_{GS} - V_T)$. The bias voltage, V_{bf}, sets the resistance of the MOS device in Figure 5.10(b).

Another consideration for bias filter design is the time constant of the filter which determines the start-up time of the VCO. If the bias filter cut-off frequency is $5KHz\,(R = 1M\Omega, C = 30pF)$, then the start-up time will be around $200\mu s$. This severely affects the PLL lock time. To speed-up the start-up of the VCO, proposed solutions are shown in Figure 5.11. A delay in power down circuit in the bias filter of Figure 5.11(a) is added to speed-up the start-up. The MOS device, M_5, is used as a switch to bypass the bias filter resistor at start-up. A delay in the bias filter of Figure 5.11(b) is implemented dynamically. The MOS device, M_5 exhibits small resistance at power-up since the node X in Figure 5.11(b) is at V_{DD} voltage level. Once M_2 and M_7 start conducting

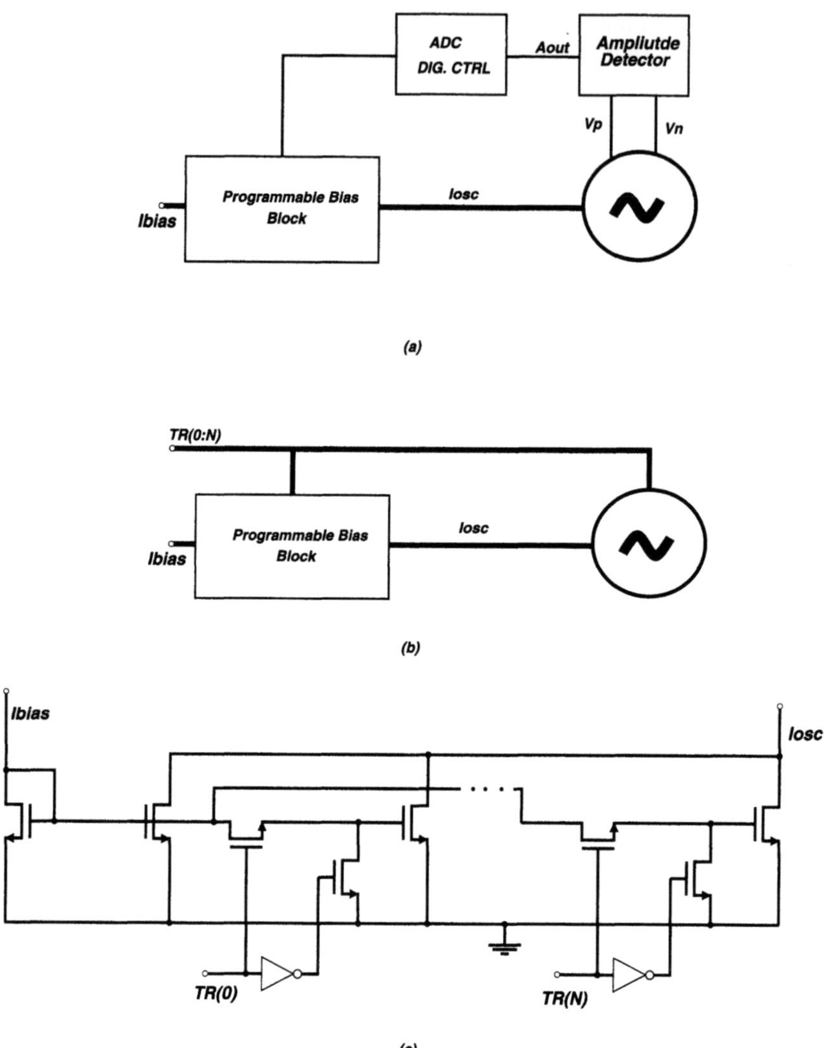

Figure 5.9.   (a) amplitude correction circuit with detector (b) amplitude correction based on tuning curve selection (c) programmable bias circuit.

current, the voltage at node X decreases from V_{DD} to V_{DS2}, and hence the resistance of M_5 increases.

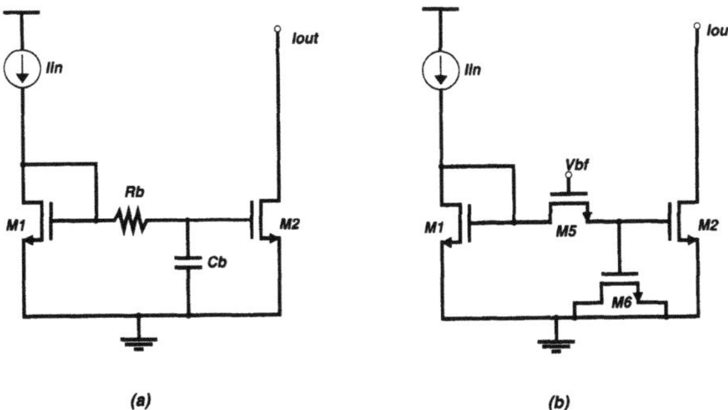

Figure 5.10. Bias filtering (a) conventional low-pass bias filter (b) proposed bias filter.

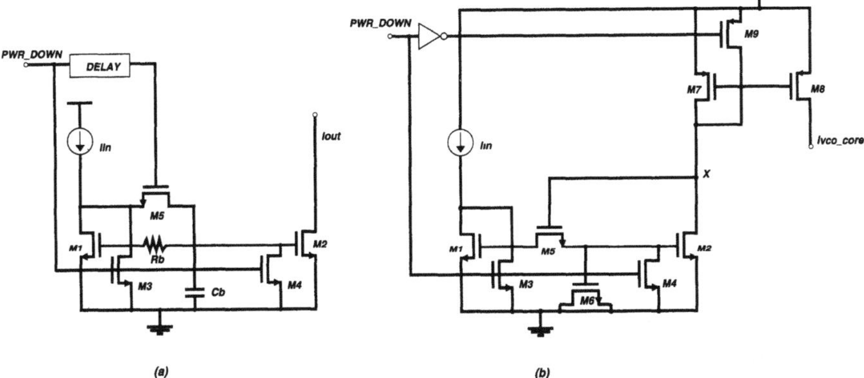

Figure 5.11. Bias filtering speed-up (a) power-up delay circuit (b) dynamic delay circuit.

4. Resonator Tank Design

The key parameter for a fully integrated VCO performance is the quality factor of on-chip inductors, capacitors,and varactors. The loaded Q of the resonator tank is primarily determined by the quality factor of on-chip inductors and varactors. This section overviews integrated inductors and varactors.

Integrated Inductors

Inductors can be implemented in three different ways in IC technology; (i) external off-chip inductors (ii) packaging bond-wires as inductors (iii) on-chip spiral inductors.

The use of external resonators (inductors or striplines) is not preferred with CMOS technology for several reasons. The pin parasitics of the package will limit the usable values. The crosstalk paths between pins will inject noise into the resonator tank and degrade the noise performance of the VCO. Also, ESD (electro-static discharge) protection networks in CMOS are probably the major factor preventing the implementation of external resonator tanks.

Although bondwire inductors have a very high quality factor, they are not commonly used in VCOs because of lack of reproducibility and mechanical stability. An excellent review of design and implementation of bondwire inductors can be found in the reference [9].

On-chip integrated inductors are favored over off-chip ones because pad and bond wire parasitics are eliminated. Also, on-chip inductors exhibit good reproducibility since the inductor value is mainly determined by horizontal dimensions which are tightly controlled by lithographic resolution in any CMOS technology. The major drawback of on-chip inductors is the low-Q factor and large die area. On-chip integrated inductors are built in spiral geometries including squares, octagons and circles as shown in Figure 5.12. Compared to a circular inductor, a square spiral inductor has larger inductance-to-area ratio but contributes more series resistance which is due to longer overall metal length for a given inductance value. Therefore, a spiral structure that more closely approximates a circle (whenever technology permits) is preferred to increase the quality factor.

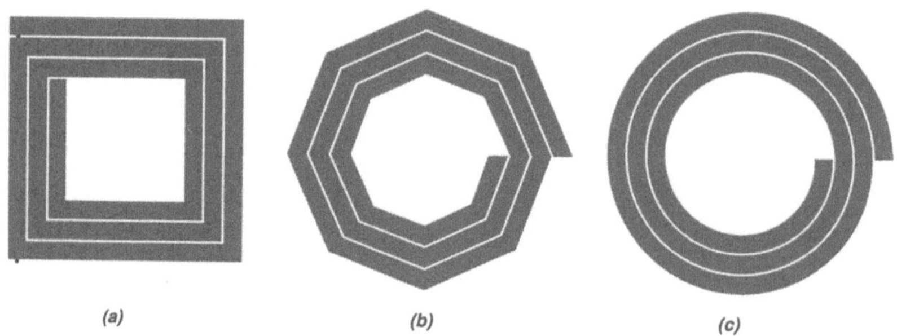

Figure 5.12.   Integrated spiral inductor geometries (a) square (b) octagonal (c) circular

The most important parameter for an integrated inductor is the quality factor. The quality factor of integrated inductors in CMOS technology suffers from three loss mechanisms; metal sheet resistance (ohmic loss), capacitive coupling to the substrate, and magnetic coupling to the substrate [7]. Approaches to reduce these losses and obtain high Q on-chip inductors are listed below;

■ Reduce metal sheet resistance by using thicker metalization [56], stacking of metal layers, and using lower resistivity metals (e.g. copper) [57].

■ Make the dielectric layer between metal layers and the substrate as thick as possible by using top metal layers.

- Reduce substrate losses by using high-resistivity substrate($\rho > 1k\Omega{\cdot}cm$) [58], by selectively removing the underlying substrate with post-fabrication steps [59], by using patterned ground shield (This method is quite useful for low-resistivity substrates, $\rho < 0.1\Omega \cdot cm$) [60].

All approaches depend on the technology parameters. Modern RF CMOS technologies offer a thick top metal layer between $2 - 10\mu$ on a medium resistivity substrate ($1\Omega \cdot cm < \rho < 20\Omega \cdot cm$).

One of the key issues in the use of an on-chip inductor in a circuit design is the adequate prediction of its behavior. A straightforward method used by CMOS foundries is to fabricate and measure a whole batch of inductors with varying geometries. A library of inductors is obtained from measurement data. This library is also extended by fitting measurement data to simple models. The fitted simple models allow only changes in one of the geometry parameters around measured inductors, and hence limiting the available inductors to a certain subset of the measured inductors. This is obviously not well suited for optimum VCO inductor design since the maximum Q and the smallest die are needed at a given frequency of interest.

Another approach is to use an electromagnetic (EM) simulator to characterize the behavior of a specific inductor structure. Commercial and free softwares are available for on-chip inductor simulations such as Agilent Momentum [61], FEMLAB [62], Sonnet [63] and ASITIC [64]. Every simulator has its own strengths and disadvantages such as speed, accuracy, user interface, technical support, etc. It is beneficial to have some familiarity with the methods used in the simulators. ASITIC (Analysis of Si Inductors and transformers for ICs) is used for inductor simulation in this book [64, 65]. ASITIC can simulate inductance L, series resistance R_L, substrate capacitance C_{S1} and C_{S2}, and substrate resistance R_{S1} and R_{S2}, and all the parameters are shown in Figure 5.13.

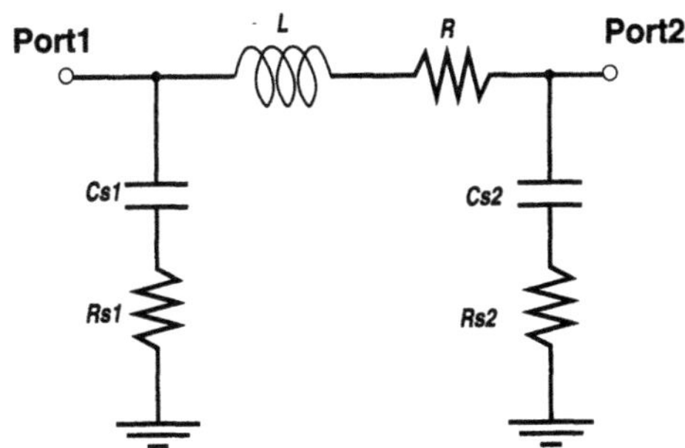

Figure 5.13. Narrow band inductor SPICE model in ASITIC.

Simple formulas for estimating the inductor value for a given geometry can be found in Reference [66]. The general relationship between an inductor and its geometrical dimensions can be expressed as follows;

$$L \propto N^2 A/l \tag{5.24}$$

$$R_L \propto N \tag{5.25}$$

where N is the number of turns, A is the cross-sectional area, and l is the total length of the spiral inductor. By connecting two layers of spiral inductors in series, inductance can be increased by 4 times with the same inductor area since inductance value is proportional to N^2. Moreover, the series resistance is proportional to N resulting in the quality factor of the inductor being improved simultaneously. However, the quality-factor of two-layer inductors is smaller than twice that of single-layer inductors since the lower layer of metal usually has a higher sheet resistance and larger substrate capacitances.To increase the inductance while maintaining reasonable quality factor, multi-layer inductors can be used [67].

Differential Inductor

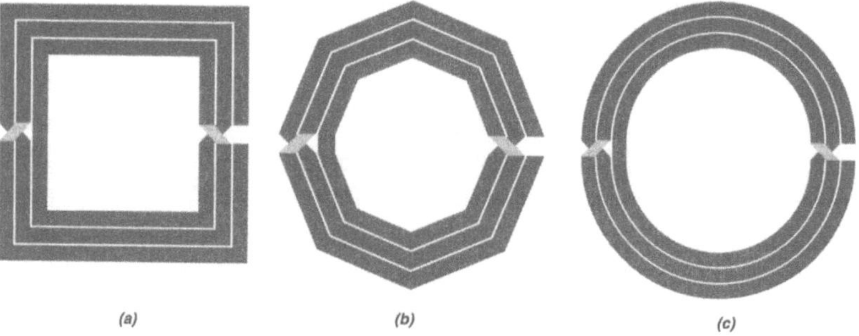

Figure 5.14.   Integrated differential spiral inductor geometries (a) square (b) octagonal (c) circular.

Spiral inductors also can be implemented in differential structure as shown in Figure 5.14. Differential inductors exhibit higher Q, better common mode rejection, and smaller die area due to smaller parasitic capacitance (about half) than two single-ended inductors [68]. A differential inductor can be viewed as two single-ended inductors inter-wounded symmetrically. A single differential spiral inductor shown in Figure 5.15(a) can be used to replace a pair of single ended inductors in the physical layout Figure 5.15(b).

Simulated model parameters for the inductors shown in Figure 5.15 are,

```
Two single ended inductors;
```

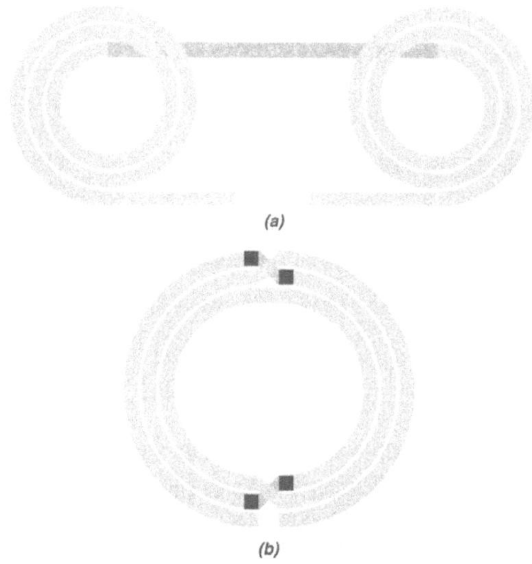

Figure 5.15. (a) two single-ended inductors used for differential operation (b) a single differential center-tapped inductor.

```
Pi Model at f=4 GHz:   Q = 7.301 , 7.361 , 8.677
L = 2.342 nH R = 5.59
Cs1 = 49.1 fF Rs1 = 684.4
Cs2 = 47.82 fF Rs2 = 723.1        Est. Resonance = 14.84 GHz


Differential inductor:
Pi Model at f=4 GHz:   Q = 10.35 , 10.33 , 12.22
L = 2.287 nH R = 3.879
Cs1 = 36.87 fF Rs1 = 830.8
Cs2 = 37.03 fF Rs2 = 842.8        Est. Resonance = 17.33 GHz
```

The differential Q factor is the third one listed for Q values, and the first two Q terms are for the single ended excitation of each of the two ports separately. The Q factors are 8.7 and 12.2 for single ended and differential structures, respectively.

Integrated Varactors

Integrated varactors in CMOS technology are important components for the integration of tunable LC tank circuits in designing RF filters and VCOs. The quality factor, Q, and the C-V characteristic are two important parameters of a varactor for VCO design consideration. The C-V characteristic is especially critical for VCO design since it has impact on the tuning range and tuning linearity. The following devices in a CMOS technology are often used as varactor [44, 45, 69];

(i) p+ to n-well junction

(ii) Accumulation Mode MOS (AMOS)

(iii) Standard Mode NMOS

(iv) Standard Mode PMOS

The first three devices are often preferred in VCO design because they use electrons as majority carriers. As the mobility of electrons is higher than the mobility of holes, the n-type varactors are expected to show higher Q than p-type ones [45].

p+ to n-well junction

The p+ to n-well junction varactor utilizes the junction capacitance C_j associated with the depletion region between the p^+-diffusion and the n-well as shown in Figure 5.16(a). The value of the junction capacitance C_j is controlled by the reverse voltage, which is defined from the Cathode (C) to the Anode (A). The varactor capacitance, C_V, can be expressed as,

$$C_V = \frac{C_{jo}}{\left(1 + \frac{V_R}{V_J}\right)^M} \tag{5.26}$$

where C_{jo} is zero-bias junction capacitance, V_R is the reverse DC voltage across the junction (V_{CA} in Figure 5.16(a)), V_J is the junction potential, and M is the grading coefficient.

Accumulation Mode MOS

The Accumulation Mode NMOS varactor is a n-channel MOS placed in an n-well as shown in Figure 5.16(b). The varactor function is achieved by changing the mode of operation from depletion to accumulation, by which the capacitance is changed from minimum to maximum. With a negative voltage applied between the gate and the drain/source, electrons just beneath the gate are pushed away and a depleted area is created, in which the total capacitance from gate to drain/source, $C_{G,D/S}$, is the series connection of the oxide capacitance, C_{ox}, and the depletion capacitance, C_d. If the voltage is reversed so that a positive voltage is applied between the gate and the drain/source, the silicon surface is accumulated with electrons from the two n^+-diffusion areas. $C_{G,D/S}$ is then obtaining its maximum value, which is equal to the oxide capacitance C_{ox}.

Standard Mode NMOS

The Standard Mode NMOS varactor is the same as the n-channel MOS in terms of layout and basic operation with the exception that the drain and

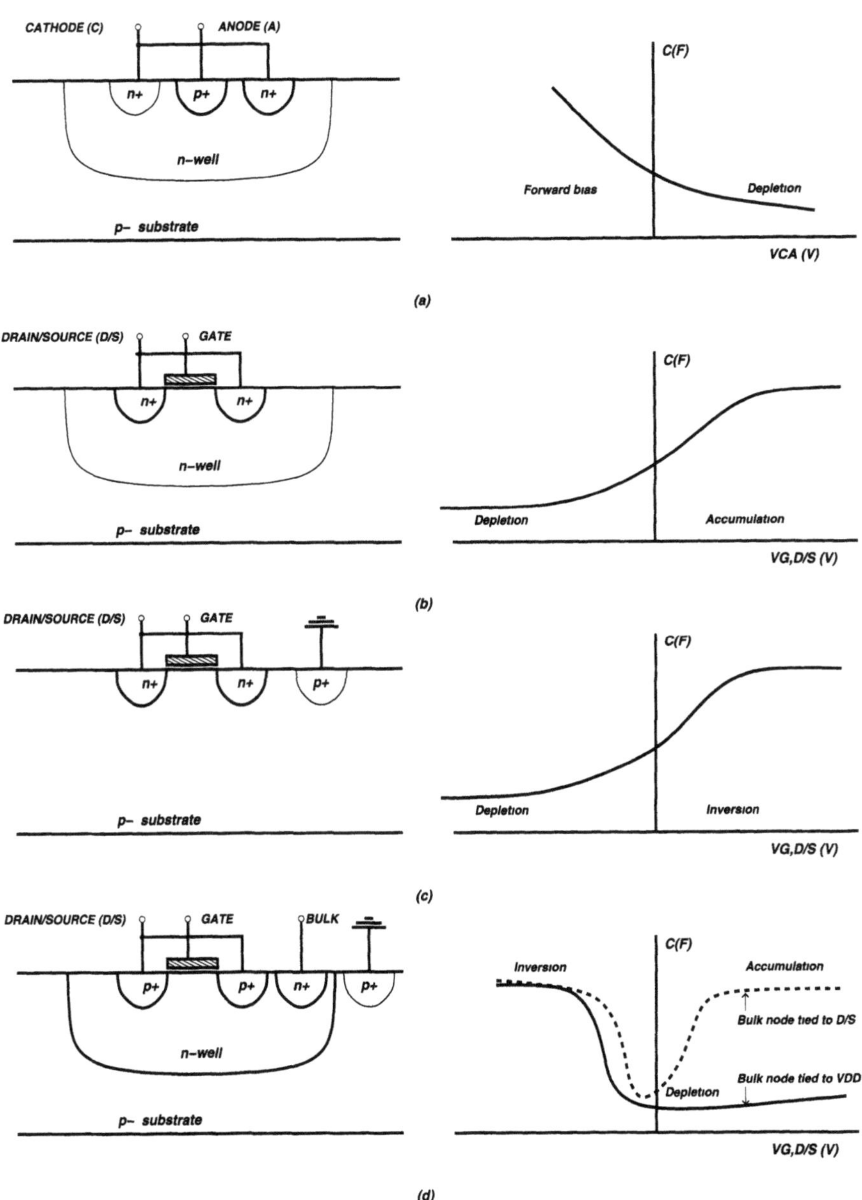

Figure 5.16. Varactor structures in CMOS technology and C-V characteristics (a) p+ to n-well junction varactor (b) NMOS varactor (c) Accumulation mode MOS varactor (d) PMOS varactor.

source are shorted to form a single terminal as shown in Figure 5.16(c). The varactor function is achieved by changing the mode of operation from depletion to inversion, by which the capacitance is changed from minimum to maximum. With a small positive voltage applied between the gate (G) and the drain/source (D/S) a depletion region is created just beneath the gate. In this situation the total capacitance from gate to drain/source, $C_{G,D/S}$, equals the series connection of the oxide capacitance C_{ox} and the depletion capacitance C_d. If the gate voltage is increased, the depletion region beneath the gate will extend deeper into the substrate, which will cause C_d and consequently $C_{G,D/S}$ to increase. When the gate voltage is increased even further, an inversion layer (channel) is created at the silicon surface. When strong inversion is reached, the $C_{G,D/S}$ will obtain its maximum value, which is equal to the oxide capacitance C_{ox}.

Standard Mode PMOS

The Standard Mode PMOS varactor is the same as the p-channel MOS in terms of layout and basic operation with the exception that the drain and source are shorted to form a single terminal as shown in Figure 5.16(d). It has third terminal because of the n-well connection, 'Bulk'. It can exhibit different C-V characteristics depending on the bulk terminal biasing. The C-V characteristics are shown in Figure 5.16(d) when the bulk is tied to drain/source (D/S) terminal or VDD. The varactor function is achieved by changing the mode of operation from inversion to depletion when the bulk is tied to VDD by which the capacitance is changed from maximum to minimum. When the bulk is tied to the D/S terminal, the mode of operation exercises inversion, depletion, and accumulation as well. With strong inversion and accumulation, the $C_{G,D/S}$ will obtain its maximum value, which is equal to the oxide capacitance C_{ox}.

The actual voltage applied across to the varactor device terminals is composed of DC voltage and large signal AC voltage. Therefore, the C-V curve calculated for DC voltage differs from the large signal added. The C-V curve under large signal condition depends on the amplitude of the signal applied across its terminals along with DC voltage. The effective capacitance is obtained through averaging the instantaneous capacitance over one oscillation period for VCO operation. A test structure to simulate the C-V characteristic of a PMOS device is shown in Figure 5.17(a) with a DC voltage applied across the terminals V_g and V_d along with sine wave with amplitude V_p. The simulated C-V curve is shown in Figure 5.17(b).

5.   Summary

In this chapter, the design and implementation of broadband VCOs at the circuit level are described. Broadband tuning is implemented with discrete and continuous tuning control mechanisms to reduce VCO tuning sensitivity, and hence reduce phase noise due to tuning control line. The discrete tuning is

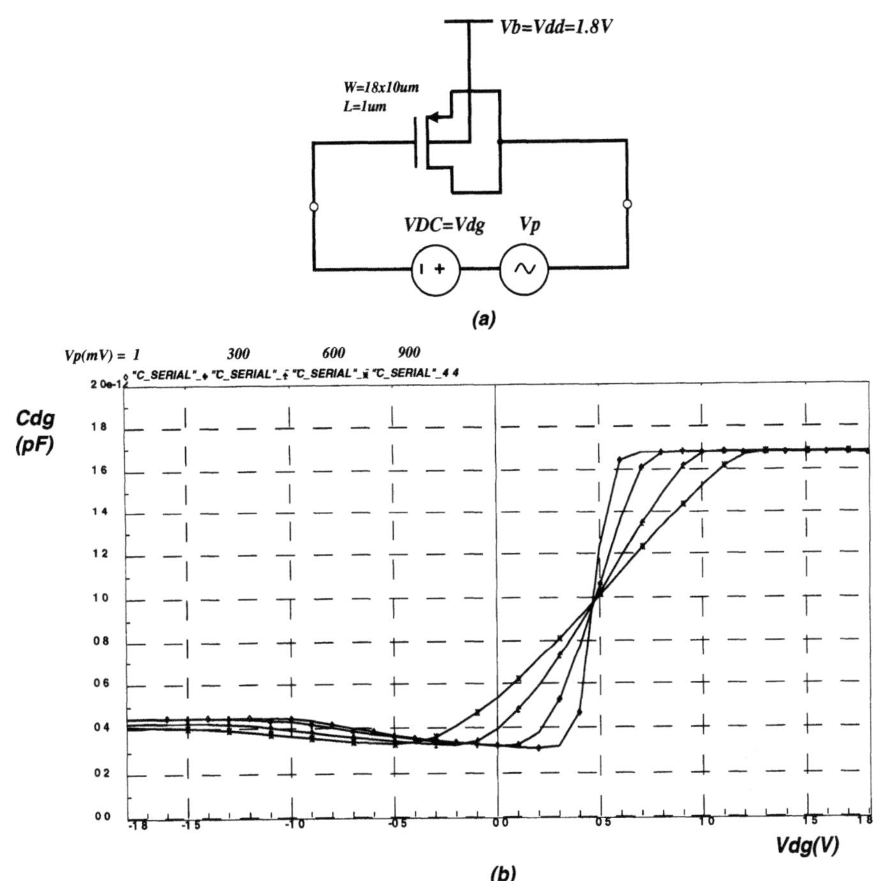

Figure 5.17. (a) test bench for simulation (b) simulated C-V curves for different AC signal amplitudes.

implemented with capacitance switching. Differential structures are favored for lower switch resistance and better common mode noise rejection. Active circuit design techniques are also discussed. The selection of a certain active circuit topology depends on the design requirements. The use of PMOS devices may lead to lower flicker noise. Integrated inductors and varactors are also described.

Chapter 6

BROADBAND VCOs: PRACTICAL DESIGN ISSUES

In the previous Chapter, the design of a broadband VCO with subbands is
presented. In this chapter, we deal with important and practical VCO design
issues due to process, supply and temperature variations, as well as VCO pulling
are discussed. Circuits and auto calibration techniques are presented for robust
VCO design. Techniques and trade-offs, at both the circuit and system levels,
fro minimizing VCO pulling are also discussions briefed.

Synthesis of a desired channel frequency with a PLL based on this broadband VCO requires assignment of a proper digital control word (trim control
word), $TR(0 : N)$ in Figure 6.1 such that the continuous control voltage,
V_{CNT} in Figure 6.1 falls into the range of desired CP output voltage range,
i.e., $V_{CP\_out,min} \leq V_{CNT} \leq V_{CP\_out,max}$ when the PLL is locked. However, the tuning curves in the subbands move up or down due to process,
supply voltage, and temperature (PVT) variations. The selection of a digital control word, $TR(0 : N)$, for each channel frequency under the condition,
$V_{CP\_out,min} \leq V_{CNT} \leq V_{CP\_out,max}$, must take into account the PVT variations. The supply voltage variation usually has less impact on the tuning range
variation since an off-chip voltage regulator and on-chip band-gap bias reference
current are used. The temperature variation has considerably less impact on the
tuning curves compared to process variations. The most important contribution
to tuning variations comes from process variations. In this chapter, the effect
of PVT variations on the tuning range are investigated, and circuit solutions are
presented to set the correct trim control word for optimum performance.

One of the single-chip radio integration issues is VCO pulling in the integrated environment. The VCO pulling occurs due to dynamic operations of
receive and transmit parts which in turn have impact on the supply and signal
lines. The VCO pulling problem is described and solutions are discussed.

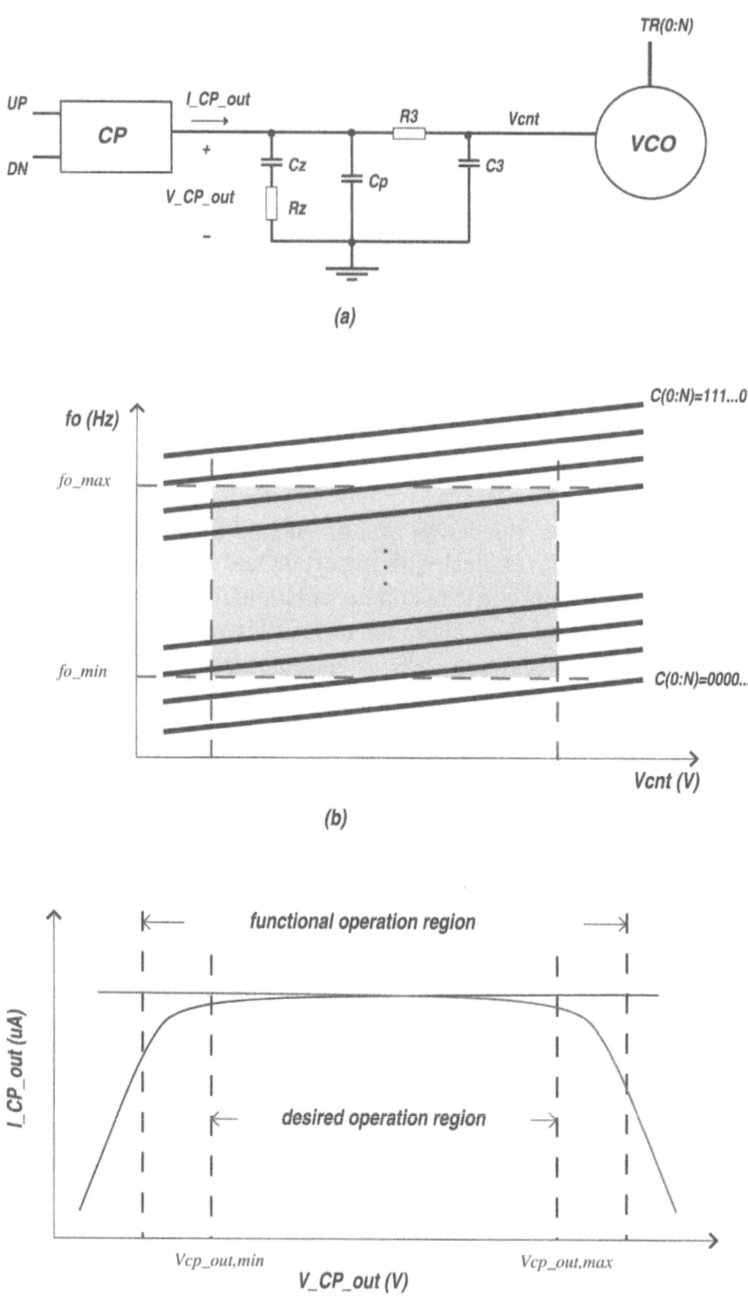

Figure 6.1.   (a) VCO with control inputs. (b) VCO tuning curves divided into subbands. (c) CP output current as a function of output voltage.

1. PVT Variation Effects on VCO Tuning

The temperature and supply voltage variations effect on the tuning range depend on the circuit topology and devices characteristics, and can be extracted from simulations as absolute maximum and minimum variations of the oscillation frequency in percent. However, the process variation has a direct impact on the absolute oscillation frequency since the oscillation frequency is equal to $1/\sqrt{LC}$. The tuning range (TR) of the VCO in terms of the tank capacitance can be expressed as (assuming typical values),

$$TR = \frac{f_{max}}{f_{min}} = \sqrt{\frac{C_{var,max} + C_{SW,max} + C_{par}}{C_{var,min} + C_{SW,min} + C_{par}}} \qquad (6.1)$$

where $C_{var,max/min}$ is the capacitance variation of the continuous tuning varactor. $C_{SW,max}$ is the maximum capacitance of switched cap for $TR(0 : N) = 111\ldots1$. $C_{SW,min}$ is the minimum capacitance of switched cap for $TR(0 : N) = 000\ldots0$. C_{par} is the fixed parasitic capacitance from the inductor, interconnects, and active device.

If we consider only the i_{th} sub-band of the VCO's entire tuning range then, the i_{th} sub-band tuning range and the tank capacitance relations can be written as follows for typical process parameters;

$$TR_i = \frac{f_{i,max}}{f_{i,min}} = \sqrt{\frac{C_{var,max} + C_{i,SW} + C_{par}}{C_{var,min} + C_{i,SW} + C_{par}}} \qquad (6.2)$$

The total tank capacitance is formed by the variable capacitance and the fixed capacitance due to the parasitics and the switched capacitance. The process variation shifts each sub-band tuning curve up or down depending on the change in the fixed tank capacitance. If we want to find the absolute frequency shift, up or down, for the i_{th} sub-band tuning range due to process variation in the capacitance; then Equation 6.2 can be re-written as;

$$TR_{i,up} = \frac{f_{i,max,up}}{f_{i,min,up}} = \sqrt{\frac{C_{var,max} + C_{i,SW,min} + C_{par,min}}{C_{var,min} + C_{i,SW,min} + C_{par,min}}} \qquad (6.3)$$

$$TR_{i,down} = \frac{f_{i,max,down}}{f_{i,min,down}} = \sqrt{\frac{C_{var,max} + C_{i,SW,max} + C_{par,max}}{C_{var,min} + C_{i,SW,max} + C_{par,max}}} \qquad (6.4)$$

For each sub-band, the absolute values of the maximum and the minimum shift can be calculated by using Equations 6.3 and 6.4 along with the process variation parameters. Typical component absolute value variations for a CMOS process are listed below.

- On chip spiral inductor value variation $\pm1\%$

- gate capacitance variation $\pm5\%$

- MIM capacitance variation $\pm15\%$

- Parasitic capacitances (inductor and interconnects) $\pm15\%$

2.   Tuning Range Calibration (Trimming)

There are two approaches to trim the tuning range of a broadband VCO with subbands. One approach is to apply the trim word externally. The other approach is the self trimming or auto calibration.

3.   External Trimming

External trimming requires the knowledge or the best estimate of trim() code from simulation or measurement. The basic steps for external trimming can be listed as;

- Determine the best trim word TR() for desired operation region from simulation or measurement (measure several chips).

- Write the determined trim word TR() from baseband at the power up.

Since external trimming doesn't correct the process variation, and therefore the continuous tuning curve must take account the process variation, this can limit the minimum VCO gain.

4.   Auto Calibration (or Self Trimming)

By auto calibration the VCO tuning range, the process variation effect to the VCO tuning range can be eliminated, and hence only effects to tuning range are to be considered; temperature and the supply voltage variations after the calibration is done. A marginal value of $\pm1\%$ can be taken for absolute VCO frequency variation due to temperature and bias variation.

Auto calibration can be done either when power-up (or reset) or while the channel frequency is changed during operation. The total lock time for PLL will be sum of the calibration time and the lock time of the PLL in the second case. The time to change from one channel to the other one is determined by the standard. There may not be enough time for auto calibration when changing channels depending on the calibration circuit and PLL architecture, and therefore auto calibration must be done at the power-up. The loop bandwidth and lock time of a PLL are inversely proportional. The loop bandwidth of a PLL is determined by the channel bandwidth for integer-N PLL architecture. Solutions for trimming at power-up and channel change are presented.

A conventional auto calibration will do the following during calibration;

- Open PLL loop from VCO control voltage line
- Set TR() code minimum or given initial estimated code

- Measure VCO output frequency for $Vmin$

- Measure VCO output frequency for $Vmax$

- If the measured VCO output frequency $fmin < fo$ for $Vmin$ and $fmax > fo$ for $Vmax$, goto the next line; or else go to next step, else TR()=TR()-1 and go back line 3.

- Keep the TR() code and close PLL loop.

Implementations of the above algorithm or similar ones do exist in the literature [70, 71].

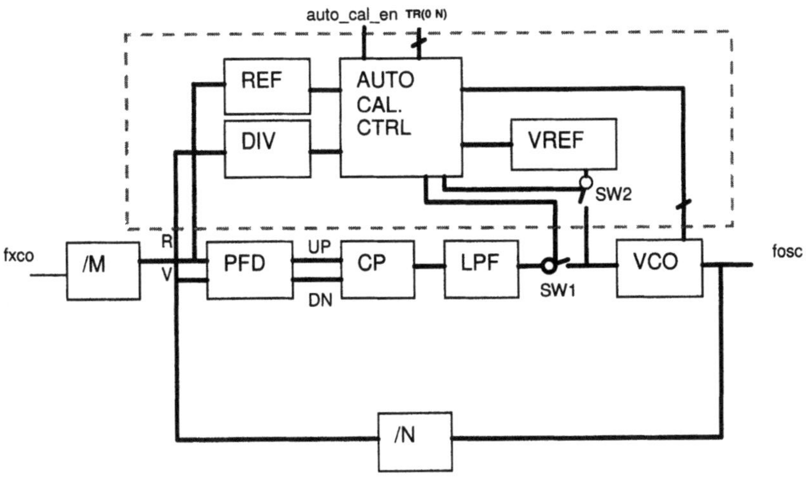

Figure 6.2.   Conventional auto calibration architecture.

Figure 6.2 shows a typical implementation of this algorithm. The PLL loop is broken with the switch $SW1$, and $Vref(Vmin, Vmax)$ is enforced by closing the switch $SW2$. The PLL divider $/N$ is set to the lower end value of $fmin$ for $Vref = Vmin$ and $fmax$ for $Vref = Vmax$. Counters, REF and DIV, will count the reference, R and divided VCO output frequencies, V, respectively. Counters will count long enough so that any possible error due to initial phases of R and V signals will be avoided. Counters, REF and DIV, will race until one of the counters finishes. If the counter REF

finishes first, then, trim code satisfies the lower edge of the tuning range, i.e., $f_o < f_{ref} \times C_{REF}$.

5. Proposed Auto Calibration Technique

We want to find the trim TRIM() code for a given VCO frequency, so that the VCO control voltage $V_{CP\_out,min} \leq V_{CNT} \leq V_{CP\_out,max}$ when the PLL is locked. The proposed calibration technique searches TRIM codes by starting from the highest TRIM code toward to the smallest one in order to find the desired code which satisfy $V_{CNT} > V_{min}$ as shown in Figure 6.3. There is no need to check the maximum value of the control voltage, i.e. the $V_{CNT} < V_{max}$ since the TRIM code search starts from the highest value. The search algorithm also can be started from the lowest TRIM code toward the highest one, and check for $V_{CNT} < V_{max}$.

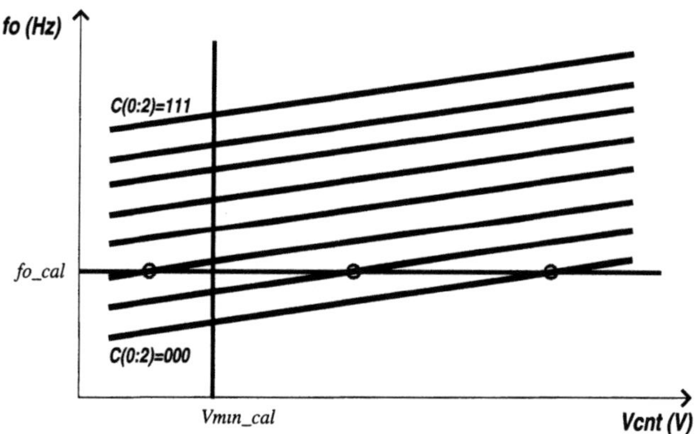

Figure 6.3.   The calibration frequency and control voltage.

An implemented architecture of the proposed calibration technique is shown in Figure 6.4. The operations flow chart of this architecture is shown in Figure 6.5. The *CALIBRATION DIGITAL CONTROL* (CDC) block controls the calibration operation. The TRIM code is set to maximum, *CAL_REQ* control signal is set to '1', and the PLL divider coefficient,*divN()*, is set to the calibrated frequency at the start-up of calibration. After setting initial values, CDS waits to allow the PLL to lock and the comparator to settle. Here, the wait time is set to the worst case value of 250usec (200usec for PLL to lock and 50usec for the comparator to settle). During the wait time, the PLL tries to lock to the calibration frequency with the assigned TRIM code. If the TRIM code is too high, the charge-pump sends DN pulses continuously, which lowers the loop filter voltage (VCO control voltage) too close to 0V. Also, the comparator compares the VCO control voltage with the reference voltage during the wait time, and reaches a final value by the end of the wait time. After the wait time is reached, CDC checks the *CAL_DONE* signal, if it is 1 then, TRIM code is set as VCO_TRIM code, and *CAL_REQ* control signal is set back to '0'. If *CAL_DONE* signal is '0', then CDC checks TRIM() code, if TRIM() code has reached '000..00', then the calibration failed to find appropriate TRIM code. If TRIM() code has not reached the '000..0', then TRIM() code is decreased, and goes back to wait period.

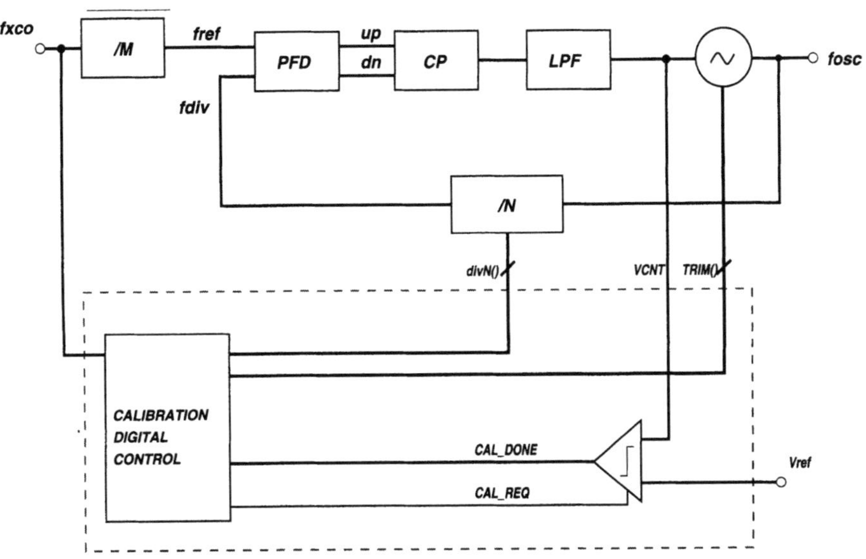

Figure 6.4.   Proposed auto calibration architecture.

CDC block is purely digital. The comparator has relaxed requirements for speed (50usec). The input offset voltage of comparator should be as small as possible to minimize the comparison error between *VCNT* and *VMIN*. The required reference voltage is obtained from the bandgap voltage reference (see Figure 6.6).

$$V_{ref} = \frac{R_1}{R_2} V_{bandgap} \tag{6.5}$$

The proposed tuning range calibration circuit is implemented in the WLAN frequency synthesizers presented in Chapter 9.

One important consideration is the total calibration time. For calibration operation, there are two approaches: i) calibrate at power up or during the reset time ii) calibrate when changing channels. The first approach is employed here. The maximum total calibration time of the proposed architecture is equal to $2^N \times W_{time}$. N is the number of the control bits. W_{time} is the wait time (250usec is used here) which depends on the worst case PLL lock time and the comparator

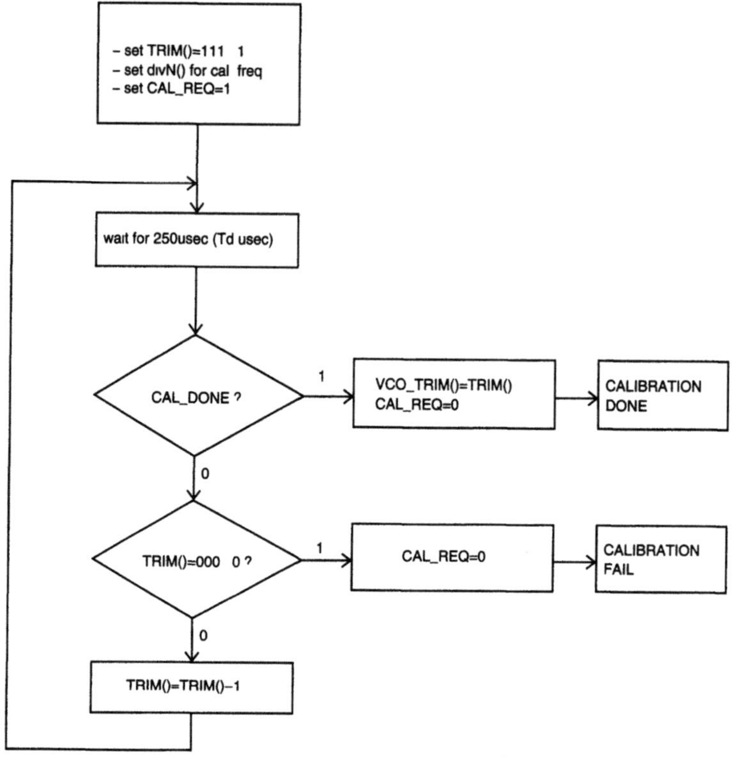

Figure 6.5.   Flow chart of operation in the proposed calibration architecture.

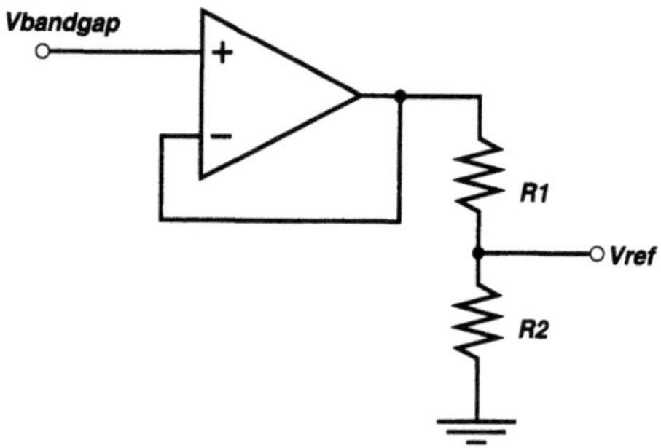

Figure 6.6.   The reference voltage circuit.

settling time. The maximum total calibration time is 2msec (8x250usec) in the worst case for the 4GHz VCO calibration. The calibration time of the proposed architecture is intended to be used at power-up. It may not satisfy the required lock time of some wireless standards if the calibration is performed during channel change. The same architecture can be modified to reduce the the total calibration time so that it can be used during channel switching. Following methods can be used for calibration to work during channel change;

- One method is to change the PLL loop bandwidth during calibration in order to make PLL bandwidth large so that PLL will acquire lock quickly (The PLL lock time is inversely related to the PLL loop bandwidth).

- A fractional-N PLL architecture can be used to decouple the PLL loop bandwidth and channel spacing (The PLL bandwidth is determined by the channel spacing for integer-N PLL architecture).

6.   VCO Pulling In the Integrated Enviroment

A single PLL frequency synthesizer is shared for the receiver and the transmitter in time-division duplexing (TDD) systems to increase component share, and hence reduce cost and power consumption since the transmitter is powered-down when receiving, and the receiver is powered-down when transmitting. Figure 6.7 shows a typical TDD radio architecture. Every wireless standard imposes rapid duplexing time between receive and transmit operation. The frequency synthesizer is always running during duplexing, while receiving or transmitting. Normally, the frequency synthesizer operation should not be affected during duplexing. Monolithic integration of receiver, transmitter, and frequency synthesizer into the same substrate has impact on the frequency synthesizer VCO operation. A rapid duplexing rate can easily throw the VCO of the frequency synthesizer out of lock. The PLL will pull back the VCO to the specified output frequency. A long recovery time than the standard specified turn-around time between RX and TX operations may prohibit proper operation of a transceiver. Therefore, eliminating "VCO pulling" or minimizing recovery time below the turn-around time specified by the standard during duplexing is necessary to integrate the frequency synthesizer along with the receiver and the transmitter into a single radio chip.

There are two aspects of this problem which must be addressed to find a solution.

- How the VCO is pulled out of lock during duplexing?

- How the PLL reacts to a VCO frequency (phase) jump when it is in phase-locked state?

These two aspects are discussed , and solutions to this problem are explored in the next sections.

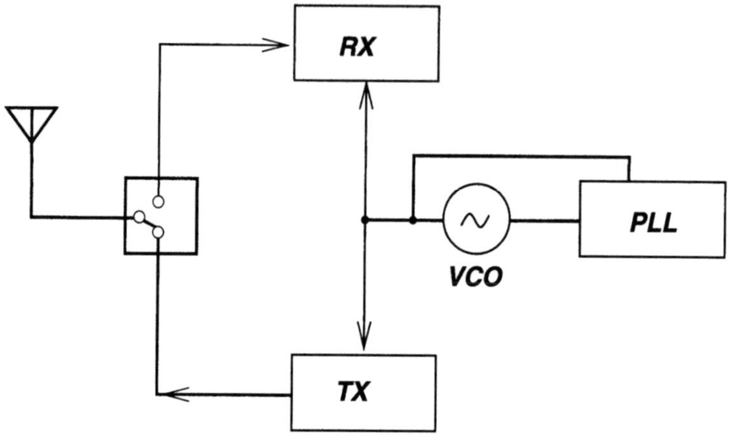

Figure 6.7.   A typical simplified TDD radio architecture.

The VCO Pulling Problem

The instantaneous oscillation frequency of an LC VCO can be written as,

$$f_{osc} = \frac{1}{\sqrt{L_{tank}[C(v) + C_{par} + C_{load}]}} \tag{6.6}$$

where $C(v)$ is the voltage dependent capacitance seen by the resonator tank, C_{par} is the sum of the parasitic capacitance from interconnects and inductor, and C_{load} is the input capacitance of the VCO buffer. The voltage dependent capacitance seen by the resonator tank is the sum of the varactor capacitance and the active device capacitance. The instantaneous oscillation frequency of an LC VCO is, in general, function of all VCO inputs, V_{cnt}, V_{supply}, and I_{bias}. An LC VCO with its input and output connection is shown in Figure 6.8. It will be useful to express the instantaneous oscillation frequency in terms of VCO input and output parameters which can be characterized with simulation and measurements. Equation 6.6 can be expressed in terms of input and output parameters as;

$$f_{osc} = f_o + K_{VCO}V_{cnt} + K_{supply}\triangle V_{supply} + K_{bias}\triangle I_{bias} + K_{load}\triangle C_{load} \tag{6.7}$$

where f_o is the free-running VCO output frequency when $V_{cnt} = 0$, $V_{supply} = VDD$, $I_{bias} = I_B$, and $C_{load} = C_L$. K_{VCO} is the VCO sensitivity to the control voltage of the VCO in MHz/V.

$$K_{VCO} = \frac{\partial f_{osc}}{\partial V_{cnt}} \tag{6.8}$$

K_{supply} is the VCO sensitivity to changes in the supply voltage of the VCO in MHz/V.

$$K_{supply} = \frac{\partial f_{osc}}{\partial V_{supply}} \tag{6.9}$$

K_{bias} is the VCO sensitivity to changes in the bias current of the VCO in MHz/μA.

$$K_{bias} = \frac{\partial f_{osc}}{\partial I_{bias}} \tag{6.10}$$

K_{load} is the VCO sensitivity to changes in the load capacitance of the VCO in MHz/fF.

$$K_{load} = \frac{\partial f_{osc}}{\partial C_{load}} \tag{6.11}$$

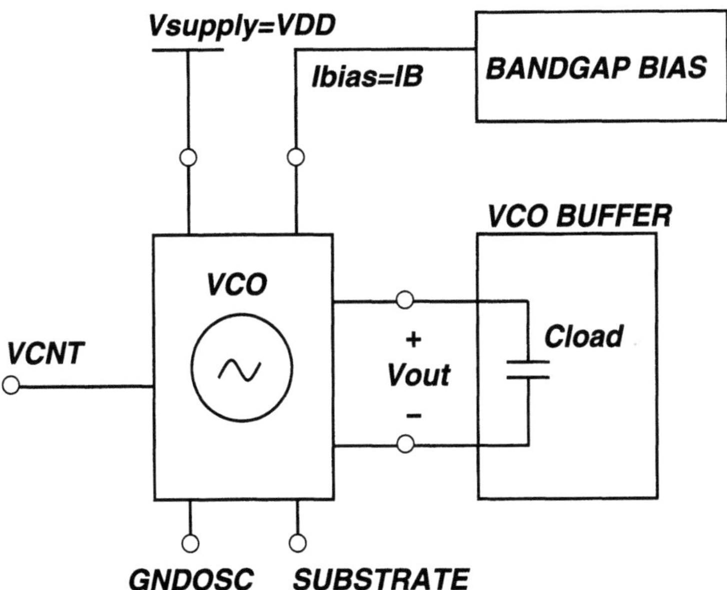

Figure 6.8. VCO with its inputs and outputs

The instantaneous oscillation frequency of a free-running VCO is difficult to predict and control, and therefore the VCO is placed in PLL loop and locked to a clean reference signal (crystal oscillator) to synthesize a desired frequency. Equation 6.7 indicates that any change in the VCO control voltage, supply voltage, bias current or load capacitance will, in effect, change the VCO output frequency. V_{cnt} in Equation 6.7 is updated by the PLL at every reference clock period. The PLL can follow and suppress any change slower than the PLL bandwidth by updating the control voltage. Therefore, static or slow variations (inside the loop bandwidth) in supply voltage, bias current or load capacitance are corrected by the PLL. The fast dynamic variations in supply voltage, bias current or load capacitance present a concern for the VCO pulling since the PLL can not track the VCO frequency for changes faster than the loop bandwidth. The dynamic changes occur in a fully integrated environment as follows;

- During RX to TX or TX to RX switching, the supply and ground bounces occur due to the fact that large amount of currents (in the order of 100mA) are switched ON and OFF by the receiver and the transmitter side. Even though, the receiver, the transmitter, and the frequency synthesizer are often placed into different supply and ground domains inside the chip, they are tied together on the PCB board through bond-wires and package leads. Bond-wires exhibit finite inductance and mutual coupling effects. The voltage drop across a bond-wire can be calculated as,

$$v(t) = L\frac{di(t)}{dt} \qquad (6.12)$$

If the transmitter is drawing 100mA, bond-wire inductance is about 2nH (L=2nH) and transmitter is powered down within 10ns, then there will be a jump in the supply/ground voltage of transmitter, $\Delta V = 20mV$. This rapid change will create bounces on the bond-wire with pad capacitance. The supply/ground bounces propagate into chip and other bond-wires by coupling effects. The supply/ground bounces occur in the receiver side as well when switching. The supply/ground bounces couple to frequency synthesizer supply.

- During RX to TX or TX to RX switching,the supply and ground bounces also propagate to the bandgap supply, causing dynamic changes at the bandgap generated reference current.

- During RX/TX switching, a load change may occur at the VCO buffer input due to finite isolation between the buffer input and output.

PLL Response to VCO Pulling

As a negative-feedback control system operating in phase variable, the PLL will try to correct the phase error at the output of the PLL phase detector if there is a change at the input or output. The PLL will, in general, respond to any phase (frequency)change at the input, output or loop path. The rate of change in the phase variable is also of interest.

The recovery time or lock time is of interest in the case of VCO pulling during the duplexing operation which causes phase (frequency) step at the PLL output. The dynamic behavior of a PLL is discussed in Section 2.0. There is a trade-off for loop performance parameters when selecting the loop bandwidth. A large PLL loop bandwidth will result in;

- Smaller lock time, and hence shorter recovery time for PLL during duplexing when the VCO is pulled out of lock.

- Better suppression of VCO noise inside the loop bandwidth.

- Less reference spur suppression.

- Larger integrated phase noise or large rms phase error

On the other hand, a smaller PLL loop bandwidth will result in;

- Longer lock time, and hence shorter recovery time for the PLL during duplexing when the VCO is pulled out of lock.

- Less suppression of the VCO noise inside the loop bandwidth.

- Better reference spur suppression.

- Smaller integrated phase noise or large RMS phase error

The transient behavior of a PLL discussed above is considered for a step change in the phase (or frequency). What if the VCO output frequency varies slowly? and, How will the PLL react to slow variation at the output or input frequencies? Any frequency variation at the VCO output frequency which slower than loop bandwidth will be suppressed by loop feedback.

Minimization of VCO Pulling

VCO pulling and PLL response to a frequency jump are over-viewed in previous two sections. Minimization of VCO pulling time involves both circuit and system level design of a PLL frequency synthesizers;

1 At the circuit levelwe find that minimizing or eliminating frequency jump during duplexing requires VCO block parameters. Equation 6.7 implies that we need to minimize all dynamic sensitivity parameters of the VCO; K_{VCO}, K_{supply}, K_{bias}, and K_{load} to minimize the VCO output frequency jump. K_{VCO} cannot be arbitrarily minimized since the VCO must have enough tuning range to cover a wireless standard required frequency band along with process, temperature, and supply variation. Supply and bias sensitivity parameters, K_{supply} and K_{bias}, of the VCO are more related to the selection of VCO topology. VCO output sensitivity to load, K_{load}, implies a buffer circuit with good reverse isolation between the VCO and mixers. A specification metric can be extracted for VCO sensitivity parameters based on radio specifications drawn from agiven wireless standard. For example, IEEE 802.11a/g standard requires RF carrier accuracy to be $\pm 20ppm$. This implies an accuracy of $\pm 48KHz$ for 2.4GHz RF carrier. Crystal oscillator used for PLL reference usually has an accuracy of $\pm 10ppm$. This implies

that the RF PLL output can tolerate frequency jump within $\pm 10 ppm$ or $\pm 24 KHz$. If the supply sensitivity of the VCO is $1MHZ/V$, then the tolerable supply jump is about 24mV. As explained in the previous section, the PLL can track any slow frequency change or variation at the VCO output. The transient effect on the supplies during duplexing between RX and TX can be reduced or minimized to a degree which can be traceable by the PLL. This can be accomplished by ramping or tapering the TX and RX supply line currents by powering up blocks in a delayed order. As shown in Figure 6.9, TX blocks are powered up in a delayed timing fashion to reduce di/dt effect.

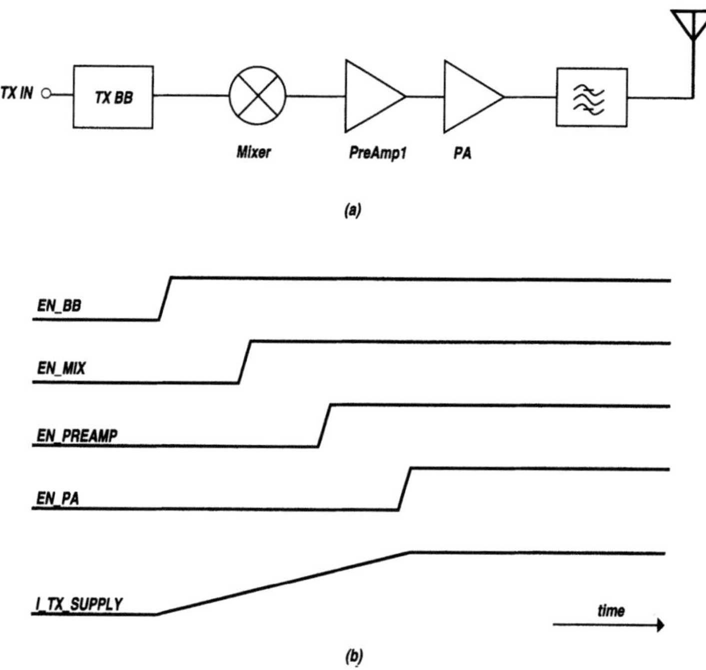

Figure 6.9.   (a) RF Transmitter (b) ramped power-up timing scheme.

2 At the PLL system level, the loop bandwidth must be optimized for the following performance parameters; i) integrated phase noise, ii) reference spur suppression, iii) lock time and iv) VCO noise suppression. The trade-off is that a wider loop bandwidth will increase the integrated phase noise and reduce attenuation of the reference clock spurs.

7.   Summary

The VCO tuning range auto calibration and VCO pulling are discussed in this chapter. The broadband tuning range is implemented with discrete and continuous tuning schemes. The discrete tuning curves can move up or down significantly with process, supply voltage and temperature variations. The conventional solution as well as a new calibration circuit are described to select the proper tuning curve for a given channel frequency.

A single PLL frequency synthesizer is shared for receiver and transmitter in time-division duplexing (TDD) systems to increase component share, and hence reduce cost and power consumption. VCO pulling occurs due to change in the VCO tuning voltage, supply voltage, bias current, etc. VCO pulling requires attention at both the circuit design level and architectural PLL system design level.

Chapter 7

4GHz BROADBAND VCO: A CASE STUDY

Design and development of a 4GHz broadband VCO in $0.18\mu m$ CMOS technology is discussed here as a case study. Potential applications for such 4GHz design have been discussed previously (Chapter 4, Figures 4.1 and 4.2). It is also used in the RF (LO1) frequency synthesizer of a multi-band multi-mode WLAN radio transceiver (see Chapter 9). The case study will focus on the VCO design for this particular WLAN application.

Complementary NMOS and PMOS and PMOS-only active circuit topologies are investigated for the VCO design. MIM capacitance and PMOS as capacitance switching techniques are explored for the band switching function. Bias filtering techniques are also investigated for noise minimization and suppression of dynamic switching noise in an integrated environment. Steady-state behavior and phase noise of VCO circuits are simulated by using EldoRF [23]. Experimental results are also presented.

1.   Design Objectives

The primary design objectives for the 4GHz VCO are to satisfy the following performance parameters: low phase noise, supply sensitivity, bias current sensitivity, and load sensitivity. Low phase noise is required to meet the integrated phase noise specification of LO1 PLL synthesizer. Low supply, bias current, and load sensitives are required for integration considerations; minimization of VCO pulling and minimization of phase noise due to noise coupling from substrate, bias, and supply lines. The tuning range requirement of the VCO is relaxed since the LO1 PLL needs to synthesize only discrete frequencies at 3840MHz and 4320MHz. An auto calibration circuit is designed to select the optimum sub-band for each frequency, and this further alleviates the broadband tuning issue. The broadband tuning issue in this design is reduced to modeling for resonator accurately by accounting for contributions from inductor, varactors, RF MOS devices, interconnects and load. Power consumption of the VCO is not a major concern since the CMOS technology used for this design has top thick metal layer (2μm thick) on a high resistivity substrate (20 Ω-cm) allowing implementation of high-Q on-chip spiral inductors which in turn leads to low power consumption.

RF modeling is the most critical aspect in this VCO design since it has an impact on the phase noise performance and tuning range. Active circuit topology selection and bias circuit design are the other critical aspects for flicker noise minimization and performance parameters consideration (supply, bias current and load sensitivities).

Circuit topologies for the VCO design are first discussed in this chapter. Implementation and simulation are also described. Measured results and conclusions are also presented.

2.   Circuit Topologies

Fully-differential complementary NMOS/PMOS and PMOS only cross-coupled topologies are evaluated for 4GHz VCO design. The complementary NMOS/PMOS VCO schematic is shown in Figure 7.1. The PMOS only VCO schematic is shown in Figure 7.2. Bias filters are used in both VCOs to suppress reference current noise and dynamic transient effects on the bias line.

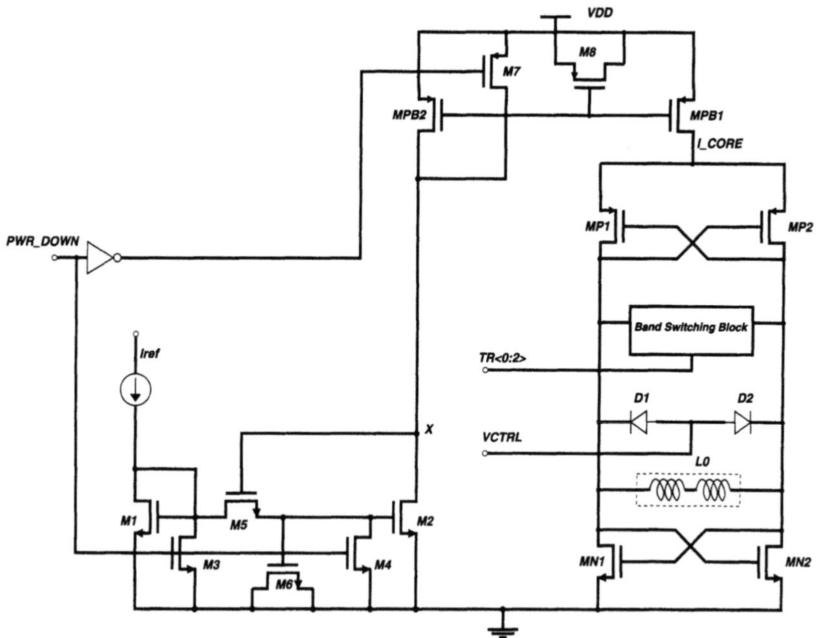

Figure 7.1.   Simplified complementary NMOS and PMOS VCO circuit schematic.

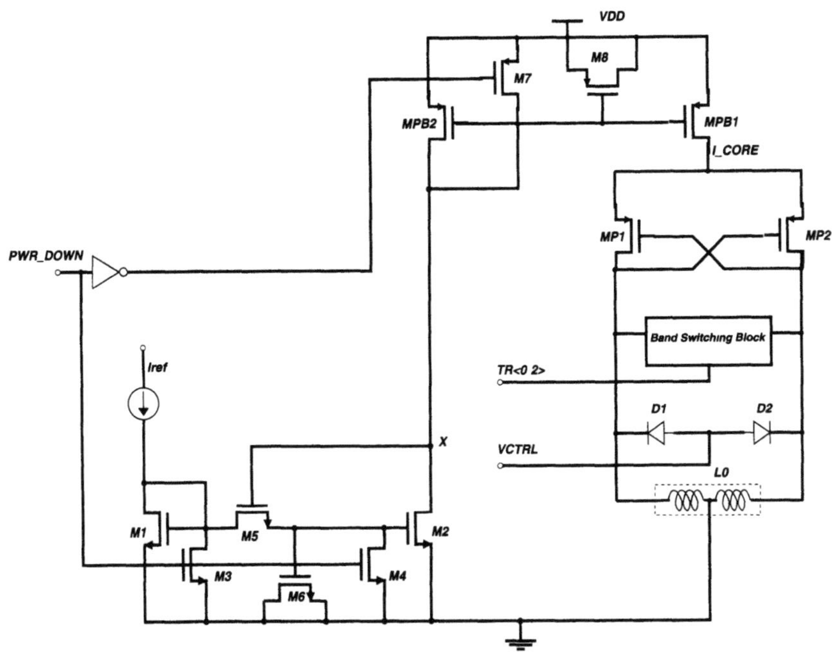

Figure 7.2. Simplified PMOS VCO circuit schematic.

Cross-coupled PMOS (MP1 and MP2) and NMOS (MN1 and MN2) pairs in the VCOs provide the transconductance to compensate for tank losses. The resonator of the VCO consists of an accumulation-mode varactor (NMOS inside n-well) for continuous tuning, a binary weighted switched capacitor array (band switching block), and a differential on-chip inductor.

The widths of the cross-coupled devices are sized to guarantee oscillation startup requirements for the minimum expected bias current while minimum lengths are used. In other words, the total small signal transconductance ($g_{m,tot}$) across the cross-coupled devices is made large enough such that the resulting initial negative resistance guarantees startup with a safety margin of at least two ($|g_{m,tot}| > 2/R_p$) under the worst-case condition. R_p is the equivalent parallel resistance of the tank.

A dynamic bias filter is employed in both VCOs since the start-up time of the VCO is important for the PLL lock time. The MOS device, M_5, exhibits small resistance at power-up since node X is at V_{DD} voltage level. Once M_2 and M_{BP2} start conducting current, the voltage at node X decreases from V_{DD} to V_{DS2}, and hence the resistance of M_5 increases as it starts operating in the triode region. M_5 acts as a resistor by operating in the triode region and M_6

acts as a MOS capacitance. Together, they form a low-pass filter to suppress noise and transients from reference current (I_{ref}).

3.   VCO Tuning

Binary weighted switched capacitor array (band switching block) is used to divide a wide tuning range into small sub bands. Two capacitance switching techniques; switching PMOS device capacitance and MIM capacitance switching using NMOS devices as RF switch are evaluated. Figure 7.3 shows the PMOS capacitance switching circuit schematic and layout. Differential layout is used to minimize interconnect resistive losses and parasitics capacitances. Figure 7.4 shows the MIM capacitance switching circuit. PMOS capacitance switching function is obtained by steering the capacitance seen at the gate of the device between depletion and strong inversion through changing the gate voltage from 0 to VDD (Chapter 5). PMOS capacitance switching offers several advantages over MIM capacitance switching; (i) higher capacitance density than MIM capacitance hence smaller die are (ii) less process variation than MIM capacitance (iii) PMOS device is standard component while MIM capacitance is optional, i.e. extra cost. Nonetheless, PMOS capacitance switching suffers from two major drawbacks limiting its use in the integrated environment. First, it converts low frequency noise on its control line to phase noise through frequency modulation since the capacitance across its terminals is voltage dependent. Second, PMOS bulk has to be connected to VDD in order to obtain a large switched maximum/minimum capacitance ratio. The voltage between the bulk node and source/drain node can vary the capacitance of the p+/n-well junction diode (between source/drain and n-well). This makes the VCO sensitive to supply voltage variations and noise through the bulk terminal.

Accumulation-mode varactor (AMOS) is used for continuous tuning of the VCOs. Accumulation-mode varactor is chosen over p+/n varactor due to the fact that it exhibits higher Q than that of a junction diode, it does not have to cope with possibility of forward biasing unlike a junction diode, and it has a higher capacitance ratio for a given tuning voltage range. The physical layout of the AMOS varactor is shown in Figure 7.5(a). The C-V characteristics and schematic symbol of the AMOS varactor are shown in Figure 7.5(b) and (c), respectively. The simulated characteristics of the AMOS varactor is shown in Figure 7.6.

A differential inductor is used in the 4GHz VCOs. The inductor test structure is shown in Figure 7.7(a). The inductor is simulated and optimized using ASITIC [64]. The simulated and measured Q of the inductor are shown in Figure 7.7(b). The simulated inductor value is 1.46nH and the measured induc-

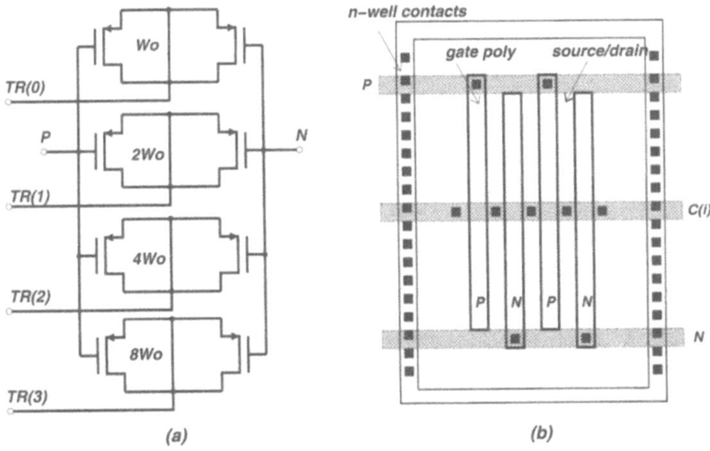

Figure 7.3. PMOS capacitance switching circuit (a) simplified schematic (b) layout.

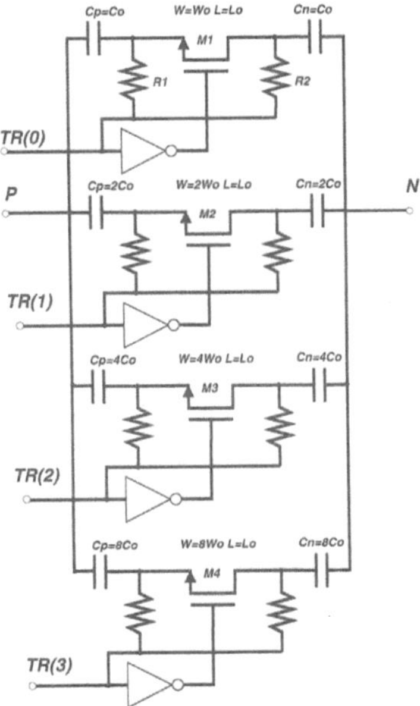

Figure 7.4. MIM capacitance switching circuit.

tance value is 1.41nH. A narrow-band model is developed for simulation using ASITIC. Figure 7.8 shows the current density of the inductor at 4GHz.

A VCO is implemented by using complementary architecture with PMOS band switching circuits. The measured results for this prototype exhibited

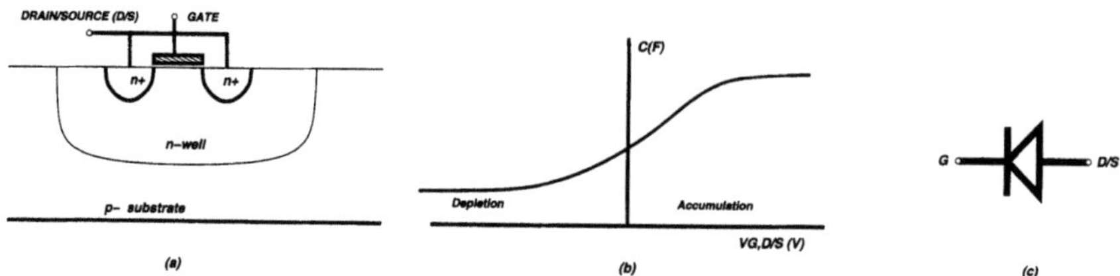

Figure 7.5.   Accumulation mode MOS varactor (a) physical cross-sectional layout (b) C-V characteristic (c) schematic symbol.

excessive supply sensitivity (150-200 MHz/V). The excessive supply sensitivity makes proper operation of the VCO impossible in an integrated environment with TDD transceiver radio. Therefore, the use of PMOS band switching is abandoned.

MIM capacitance switching is preferred and used for VCO band switching operation. Two VCO circuits with MIM capacitance switching are implemented. VCO1 employs the complementary NMOS/PMOS topology, and VCO2 employs PMOS only topology. The same tank components are used in both VCOs, i.e., same inductor and varactor. The MIM band switching capacitance sizes are adjusted to account for the cross-coupled transistor capacitances contribution to the tank.

4.   Characterization and Measurement Results

Measured tuning curves for VCO1 and VCO2 are shown in Figure 7.9 and 7.10 for different trim codes, respectively. Both VCOs use the same frequency tuning circuits, i.e., same varactor, inductor and band switching circuit. We note that VCO1 exhibits wider tuning range for same tuning voltage than VCO2, and VCO1 tuning curve is also more non-linear than VCO2 tuning curve. Tuning curves of VCO1 and VCO2 for trim code 111 are plotted together in Figure 7.11.

The differences in tuning range and linearity come from C-V characteristics of the varactor and active circuit topology. The C-V characteristic of the varactor is shown in Figure 7.6(a). The positive terminal of the varactor is biased at 0V (DC level) in VCO1, and hence VCO1 only uses C-V characteristic of the varactor over the tuning voltage range of $-1.8V < Vtune < 0$ ($V_{tune} = -V_{ctrl}$). On the other hand, the positive terminal of the varactor is biased at V_{DS2} of the NMOS device (DC level) in VCO2, and hence VCO2 uses C-V characteristic of the varactor over the tuning voltage range of $-1.8V + V_{DS2} < Vtune < V_{DS2}$. The C-V curve is the steepest at/around 0V. Therefore, continuous VCO1 tun-

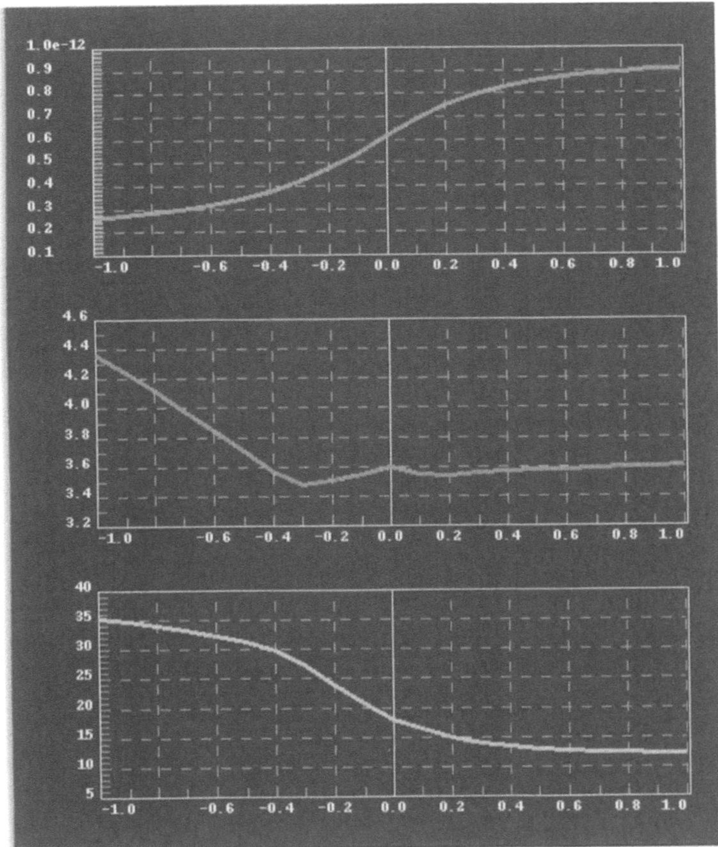

Figure 7.6. Simulated characteristics of accumulation mode varactor (a) C-V characteristics (b) Series resistance versus tune voltage (c) Q versus tune voltage.

ing curves exhibit wider tuning range and more non-linearity than that of the VCO2.

Phase noise plots of VCO1 are shown for $TR = 000, Vcnt = 1.7V$ and $TR = 111, Vcnt = 0.7V$ in Figure 7.12(a). The output frequency is 3840MHz for both settings, i.e., the LO1 PLL can synthesize 3840MHZ from different bands with different control voltages. Also, 3840MHz frequency is shown as horizontal line in Figure 7.9. Simulation results and measurement indicate that

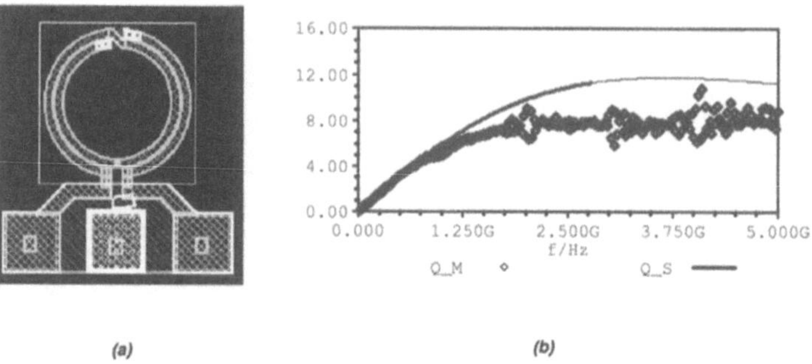

(a)   (b)

Figure 7.7.   (a) test structure of the differential inductor (b)comparison of measured and simulated Q of inductor.

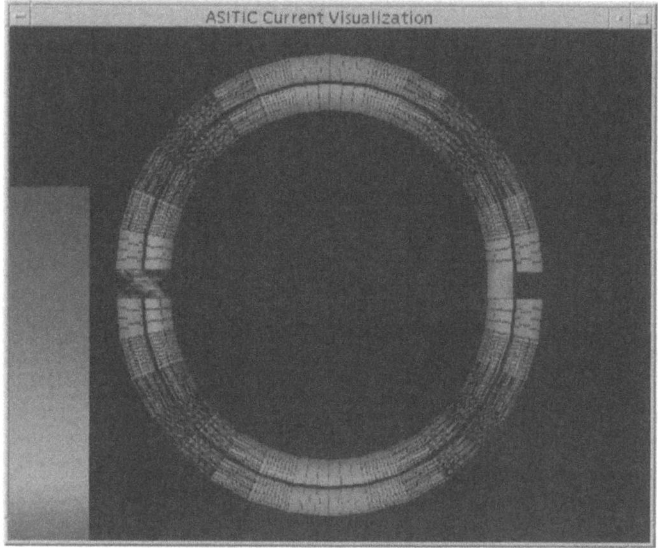

Figure 7.8.   Current density on the differential inductor at 4GHz.

the phase noise of VCO1 varies 10 dB or more with control voltage within a sub-tuning band. The phase noise variation due to control voltage can be attributed to the varactor characteristics. The non-linear C-V characteristics of the varactor (Figure 7.6(a)) exhibits steep transition for control voltages, $|V_{ctrl}| < 0.3$. This steep region occurs in the tuning curve of the VCO (Figure 7.9) in the control voltage range of $0.5 < V_{ctrl} < 1.2$ since the positive terminal of the varactor is biased at $V_{DS2} = 0.8V$. The non-linearity of C-V curve cause significant variations in the VCO sensitivity over the tuning voltage range (80MHz/V -

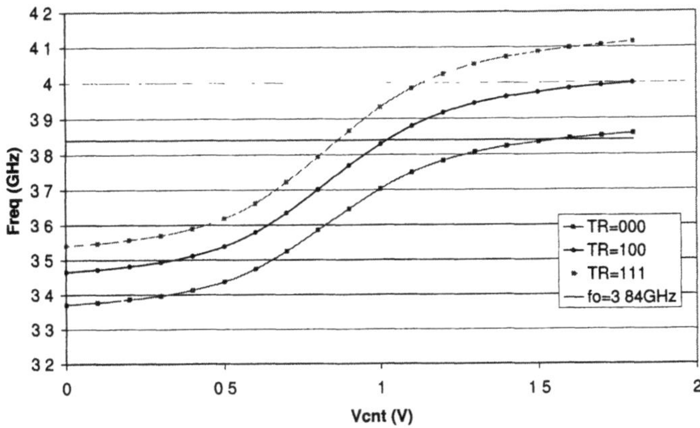

Figure 7.9. Measured tuning curve of VCO1 for different trim inputs.

800MHz/V). Another varactor characteristic is the quality factor. The Q of the varactor is shown in Figure 7.6(c). The Q of the varactor drops to half of its maximum Q value for a tuning voltage of 0 V or less. This corresponds to the VCO control voltage of $V_{DS2} = 0.8V$. The VCO resonator loaded Q is degraded by the varactor for control voltages of $V_{DS2} = 0.8V$ or lower. Also, the phase noise varies 2 dB or less for different bands while keeping the control voltage constant. This can be attributed to switching MIM capacitance circuit Q which varies with its control inputs. The Q of the switched capacitance band is maximum when all switches are off (TR=111). It reaches its minimum value for 000 control setting since all ON switch resistances are added to MIM capacitances in series.

The phase noise plot of VCO2 is shown for $TR = 111, V_{ctrl} = 0.5V$ in Figure 7.12(b). The phase noise of VCO2 varies 1 dB or less with control voltage within a sub-tuning band for control voltages of 0.4V or greater. This can be attributed to biasing of the varactor in VCO2.

Measured supply sensitives for VCO1 and VCO2 are shown in Figure 7.13. The y-axis in Figure 7.13 shows the difference between the oscillation frequency at the nominal supply voltage (1.8V) and at the varied supply voltage. VCO1 exhibits higher supply sensitivity than VCO2. The supply sensitivity of VCO1 mainly occurs due to DC bias change of the varactor and PMOS cross-coupled

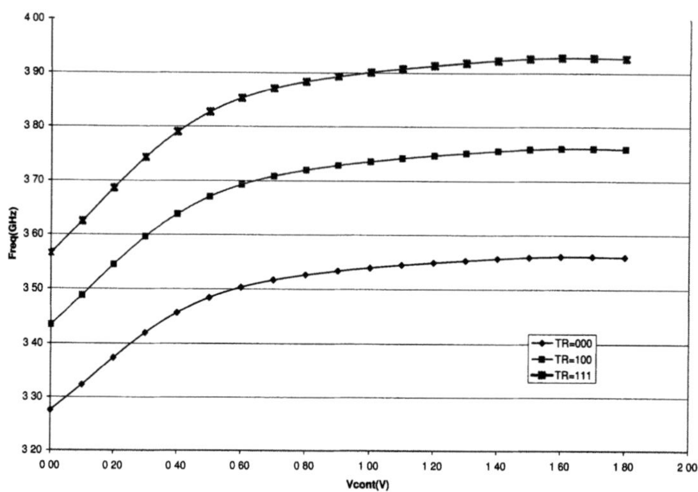

Figure 7.10.   Measured tuning curve of VCO2 for different trim inputs.

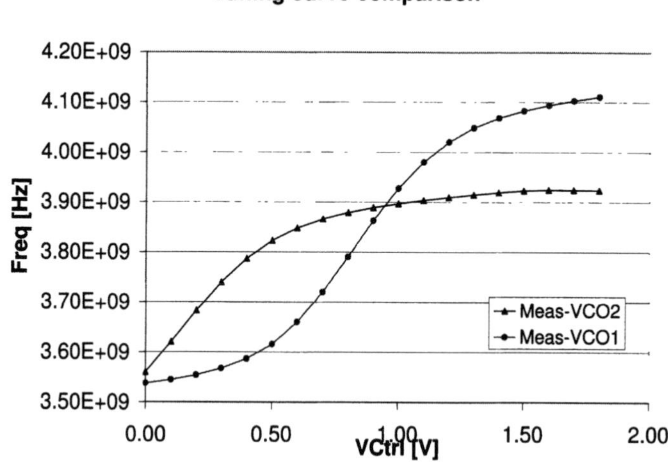

Figure 7.11.   Comparison of tuning curves for VCO1 and VCO2.

Parameter	VCO1	VCO2
Phase Noise	-	+
Supply Sensitivity	-	+
Flicker Noise	-	+
Tuning Range	+	-
Tuning Linearity	-	+
Power Diss.	+	-

Table 7.1. Comparison of VCO1 and VCO2 performance parameters.

n-well node bias change. However, the supply sensitivity of VCO2 occurs only due to PMOS cross-coupled n-well node bias change since the varactor positive terminal is biased at ground voltage.

Overall, VCO2 has better performance parameters than VCO1 due mainly to varactor biasing. The topology of VCO2 is used in the final implementation of the LO1 and LO2 synthesizers of the WLAN radio (see Chapter 9). Table 7.1 shows comparison of VCO1 and VCO2 performance parameters. The die photo of the first 4GHz VCO in the PLL test prototype is shown in Figure 7.14. It occupies $0.48mm^2$ including VCO buffer.

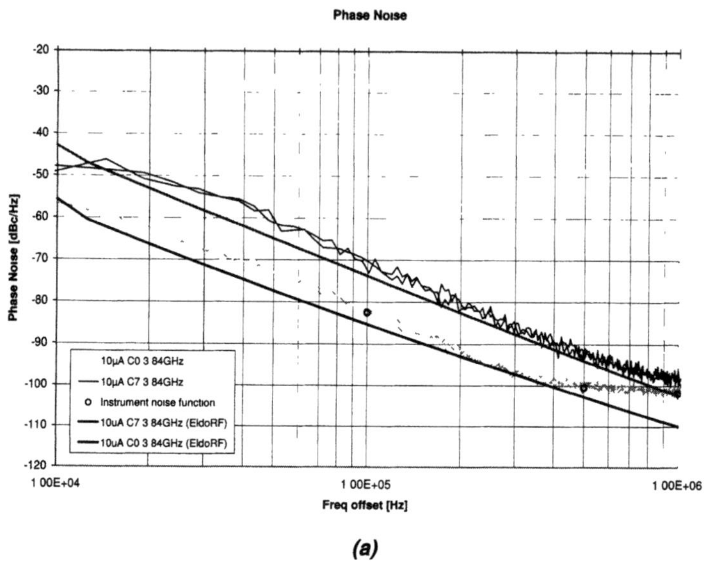

(a)

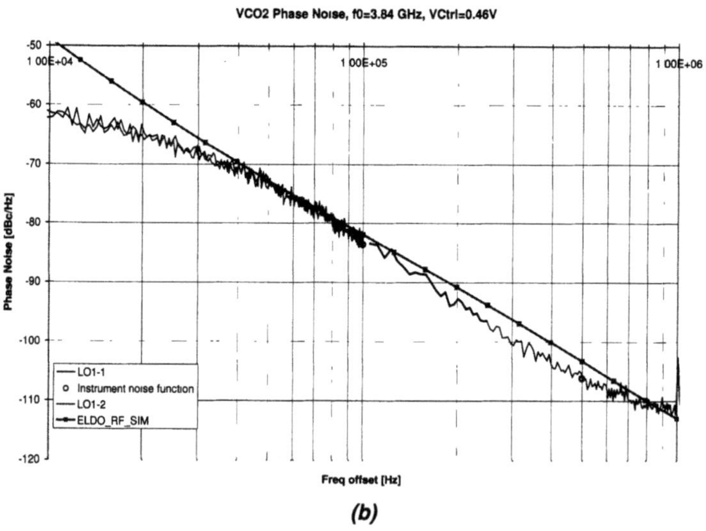

(b)

Figure 7.12.   Measured and simulated phase noises (a) VCO1 (b) VCO2.

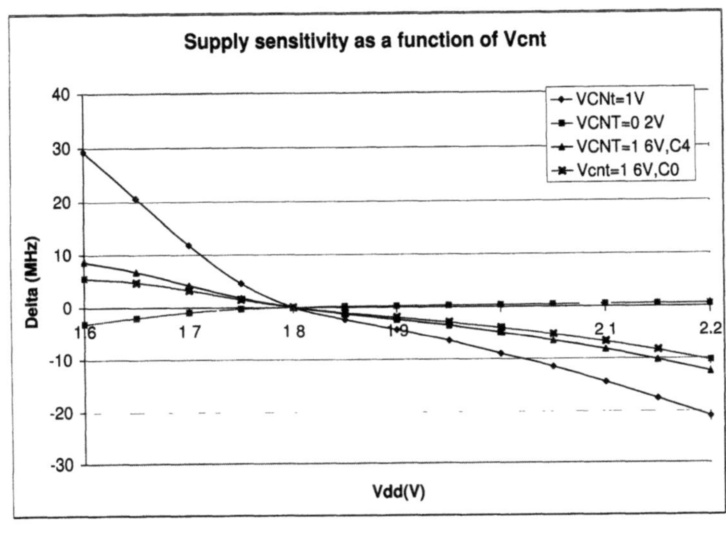

(a)

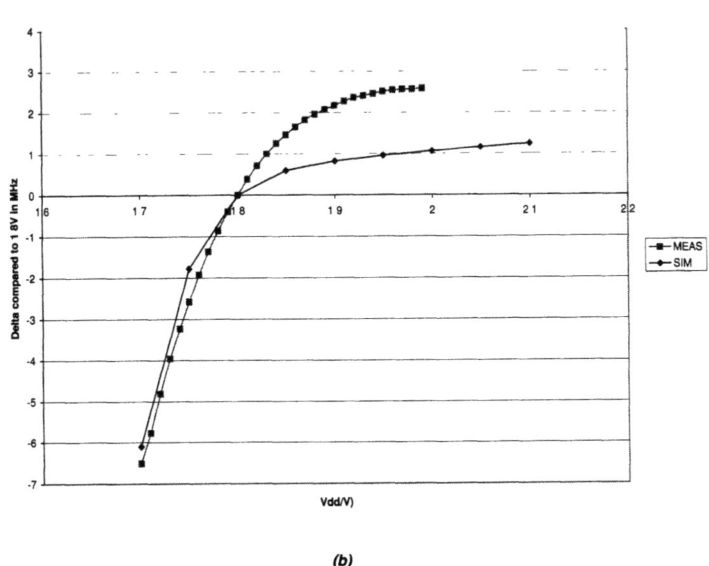

(b)

Figure 7.13. Measured supply sensitivities of (a) VCO1 (b) VCO2.

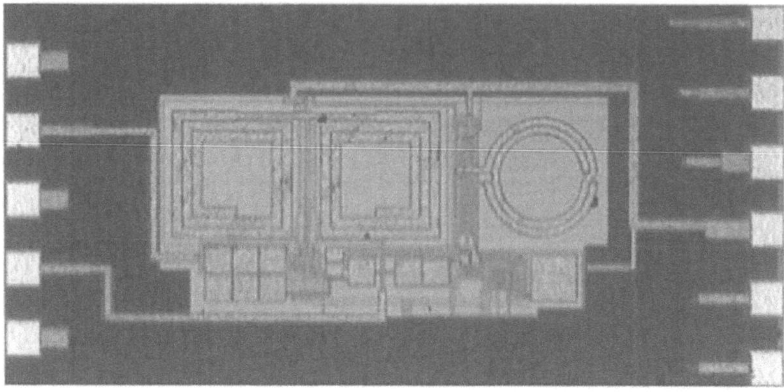

Figure 7.14.   Die photo of the 4GHz VCO in the PLL test prototype.

5.   Summary

In this chapter, design and development of a 4GHz broadband VCO in 0.18μm CMOS technology as a case study for use in the RF (LO1) frequency synthesizer of WLAN radio transceiver (Chapter 9) are presented. Complementary NMOS and PMOS as well as PMOS-only active circuit topologies are evaluated. MIM capacitance and PMOS devices are investigated for band switching. Bias filtering techniques are also investigatedfor noise minimization.

The PMOS only VCO topology exhibits better performance parameters than the CMOS topology. The better performance of the PMOS topology is due to lower flicker noise and biasing of the varactor circuit. MIM capacitance switching which exhibits better supply sensitivity is preferred over MOS capacitance switching.

Chapter 8

A PLL FOR GSM/WCDMA

The most widely used cellular standard, GSM, is coming to its bandwidth limits in big cities. In addition to bandwidth limitation, it has a low data throughput which can only handle voice communication and simple Internet access. The Wideband Code Division Multiple Access (WCDMA) standard is a candidate to replace 2nd generation standard (2G) GSM in the near future as a 3rd generation (3G) solution having increased available bandwidth and throughput for voice, data and video transmission. The transition to WCDMA (3G) from GSM (2G) requires backward compatibility, and hence this mandates a cost efficient dual-mode solution for smooth migration.

The component count in RF board design of mobile hand-held devices decreased from 500 in early 1990s to less than 150 in todays implementation [72]. The VCO is still a bottleneck in the design and high volume production of mobile hand-held devices. Early mobile devices used discrete VCO designs until very low noise and low power consumption (18-20mW) VCO modules [73] became available. The VCO module eases the design problem, but it is still an expensive solution and occupies large PC board space. Design of fully integrated VCOs is of great interest to ease these problems. A multi-band CMOS RF VCO design in a dual-mode frequency synthesizer [33] is presented in this chapter.

1. Frequency Plan and System architecture

A good frequency plan is crucial to achieving all the specifications of different standards, with a minimal amount of hardware and power consumption. The frequency plan determines how the frequency translation of the carrier is performed in both the receive and transmit paths. Therefore, the frequency plan and the synthesizer architecture should be designed concurrently to maximize hardware share for multi-standard operation. The frequency synthesizer

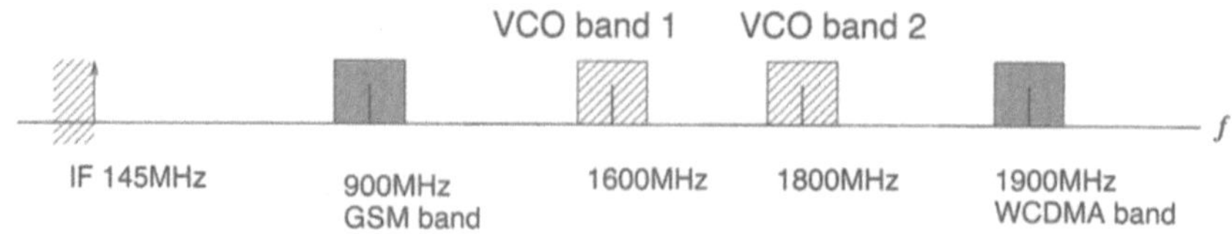

Figure 8.1. Proposed GSM/WCDMA frequency synthesizer plan for wide band IF double conversion operation.

architecture and frequency plan have a significant impact on the overall synthesizer performance, namely, phase noise, spurious tones and the required power consumption.

The frequency synthesizer is designed to generate the RF local oscillator signal for a double-conversion heterodyne radio architecture. The IF frequency of radio is chosen so that a SAW filter can be used for IF filtering.Choosing the intermediate frequency (IF) as 145MHz, the RF LO frequency range falls into 1785-1845MHz for WCDMA and 1580-1630MHz for GSM. The GSM/WCDMA receiver synthesizer frequency plan is shown in Figure 8.1.

The specifications of GSM and WCDMA standards pertaining to frequency planning are listed in Table 8.1. It can be seen from Table 8.1 that the operating frequency band and channel spacing are quite different. Hence, a dual band synthesizer for both standards requires a very wide band VCO to cover the frequency bands and quite different PLL structures due to the required channel spacing. The goal of the design is to maximize hardware share between the two modes. A dual band VCO is employed with bands; 1600MHz and 1800MHZ as shown in Figure 8.1. Also, a divide-by-2 circuit is used to generate GSM LO signal. The channel spacings for WCDMA and GSM are 5MHz and 200kHz, respectively. This would lead to different frequency divide ratio and loop bandwidth for the PLL if an integer frequency divider was used for both modes [33]. To solve problem, a fractional-N divider is employed in GSM mode with a reference frequency of 3.2MHz instead of 200KHz [33]. An integer-N divider is used for WCDMA with a reference frequency of 5MHz. This way, the reference frequency and divide ratio are close for the two modes, and allow to use the charge pump and loop filter for two modes. The system level block diagram of the proposed dual-mode synthesizer is shown in Figure 8.2.

The channel spacings are 5MHz and 200KHz for WCDMA and GSM standards, respectively. This large difference in channel spacing between the standards reflects into feedback divide ratio for PLL. To maximize hardware, WCDMA uses an integer-N architecture while GSM mode employs fractional-N architecture.

	Access scheme	Duplex approach	Frequency range (MHz)	Channel spacing	Number of channels	Modulation scheme
GSM	TDMA	FDD	890-915 (Tx) 935-960 (Rx)	200kHz	124	GSMK
WCDMA	CDMA	FDD	1850-1910 (Tx) 1930-1990 (Rx)	5MHz	12	QPSK

Table 8.1. Specifications of interest in GSM and CWDMA standards for synthesizer frequency planning.

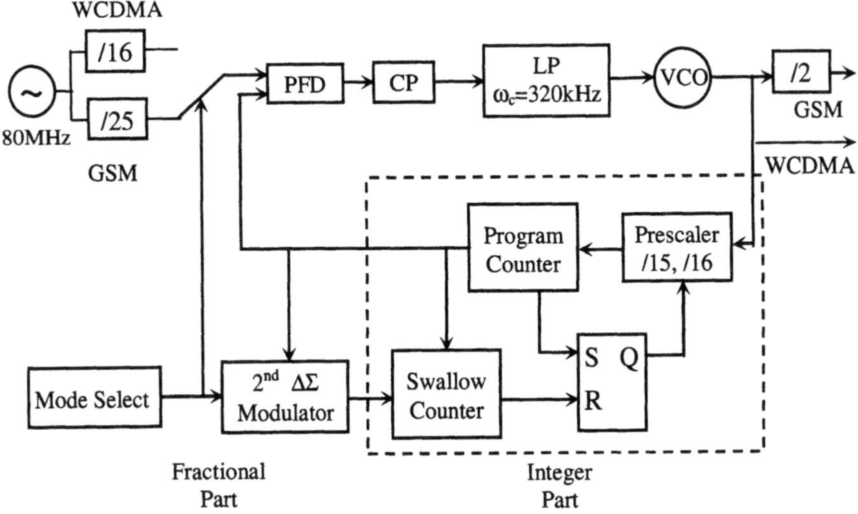

Figure 8.2. Frequency synthesizer architecture.

Closed-loop simulation is performed for loop parameter adjustment before finalizing the design for layout. A typical transient response of the VCO control voltage in the closed-loop simulation is shown in Figure 8.3.

2. Dual band VCO Design for Dual Mode Operation

A dual band LC tuned VCO is designed using integrated varactors and spiral inductors. Dual-band operation is accomplished by tuning the VCO with two control lines, one for continuous tuning and the other for digital band selection as shown in Fig. 8.4 where C_{SW} implies digital tuning and C_{FINE} implies continuous tuning. The continuous tuning is used for PLL control and channel select. The digital tuning is used for RF band selection. The active circuit of

band	Shared components				Non-shared components
	VCO	Integer frequency divider	PFD, CP, loop filter	Crystal oscillator	
GSM	1580-1630MHz	Divided by 246-254	Loop filter $f_{3dB} \sim$ 320kHz	80MHz	/2 output divider, / 32 input divider, 2nd $\Delta\Sigma$ modulator
WCDMA	1785-1845MHz	Divided by 357-369			/25 input divider

Table 8.2.   Summary of the dual-mode synthesizer configuration.

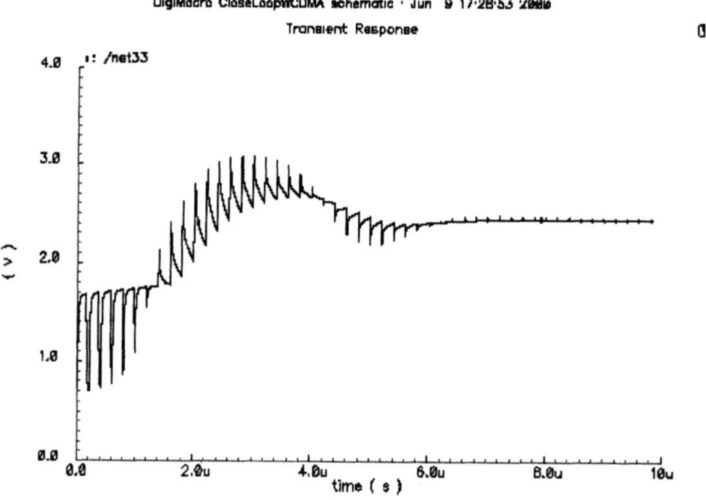

Figure 8.3.   Simulation result of the VCO control voltage transient response.

the VCO is formed by a cross-coupled NMOS transistor pair to form a negative resistance [11]. The VCO core circuit topology is similar to the ones presented in [9, 10, 34] as shown in Fig. 8.4.

The continuous tuning scheme uses an NMOS device as a varactor for analog tuning with control voltage provided by the loop filter. The digital tuning is done using PMOS transistors by operating only in strong inversion and depletion mode while the mode selection voltage takes only 0 and VDD values. In both of these modes of operation, the PMOS transistor capacitance has high quality

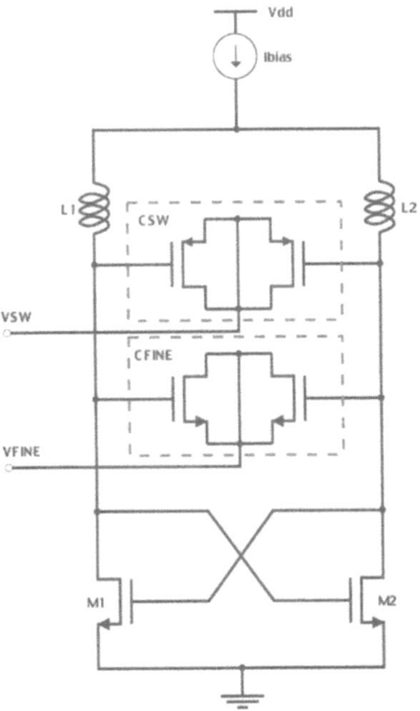

Figure 8.4.   The simplified dual-band VCO schematic view.

(Q) factor. The VCO circuit is simulated by using *SpectreRF* [24]. Simulated phase noise plot is shown in Fig. 8.5. The phase noise is about -118 dBc/Hz at 600 kHz frequency offset. The tuning curves for both bands are shown in Fig. 8.6. The VCO performance parameters are summarized in the Table 8.4.

The tuning range for a given control voltage (VFINE) is extended to cover deviations resulting from component value variations. For inductors and capacitors with a 10% tolerance the worst case of 3% frequency variation may result. The VCO gain is also taken into consideration when extending the tuning range. A rough gain, 80-120 MHz/V, for the VCO is considered for determining the loop filter components and charge pump current. The simulated tuning ranges of the VCO for both bands are shown in Figure 8.6. The VCO gain is about 100MHz/ V for both bands in the VFINE range; 0.5-2.5 V. The simulated tuning range for GSM mode is between 1500MHz-1730MHz for VFINE range; 0.5-2.5 V (VSW=0V). Also, the simulated tuning range for WCDMA mode is between 1580MHz-1860MHz for VFINE range; 0.5-2.5 V (VSW=3.3V).The desired VCO tuning range specification for dual mode operation is given in Table 8.3.

The integrated spiral inductor is designed by using top metal layer (M3) with a number of turns, N=3, metal width, W=12um, spacing, S=1.8um, and the

Mode	Frequency	VSW	VFINE
GSM	1580- 1630 MHz	0	0.5-2.5
WCDMA	1785-1845 MHz	3.3	0.5-2.5

Table 8.3. Required tuning range specifications.

Parameter	Value	Unit
Supply Voltage	3.3	V
Current	5	mA
Tuning range	1500-1850	MHz
Gain	100	MHz/V
Phase Noise at 600KHz	-118	dBc/Hz
Technology	0.5μm CMOS	

Table 8.4. Summary of VCO performance.

outer dimension, D=180um. The calculated inductor value for this geometry is found to be 2.03nH with the estimated quality factor of 4. The designed inductor contributes a 427fF parasitic capacitance to the resonant tank.

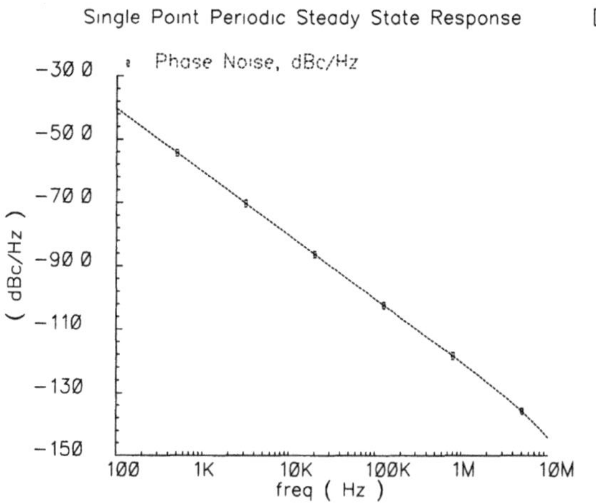

Figure 8.5. The simulated phase noise of the VCO.

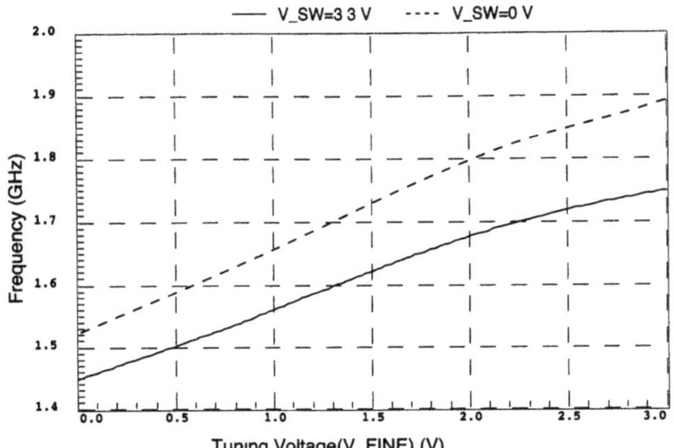

Figure 8.6.   Tuning versus control voltage simulated results for each mode.

3. Integer-N Architecture

The WCDMA mode employs an integer-N divider architecture to generate output frequencies. This architecture is appropriate for WCDMA since the channel spacing (5 MHz) in WCDMA is very large which leads to smaller divide ratios [35, 36]. For the chosen IF frequency (IF = 145 MHz), the required divide ratios are 357-369. The reference frequency is 5MHz. The simplified integer-N divider is shown in Figure 8.7. The operation of the circuit is described briefly here. Assume the program counter (PC0-PC3) and swallow counter (SC0-SC3) are set to some division ratio.When the circuit begins from the reset state (Q = 1), the prescaler divides by (N-1) until the Swallow Counter (SC) is full then the RS latch is reset, i.e. Q=0. When Q gets to 0 value, it changes the prescaler modulus to N and disables the swallow counter. The division continues on the Program Counter till it is full, then the RS latch is set high, i.e, Q=1 which in turn starts a new cycle. fout can be written in terms of fin as,

$$f_{out} = \frac{f_{in}}{(N-1)S - N(P-S)} = \frac{f_{in}}{PN - S} \qquad (8.1)$$

The division ratios for GSM and WCDMA can be obtained by setting P(program counter) and S (swallow counter) such that;

- GSM division ratio = 246 = 16x16 -10 = PxN-S → P=16, S=10 and N=16
  division ratio = 255 = 16x16 - 1 = PxN-S → P=16, S=1 and N=16.

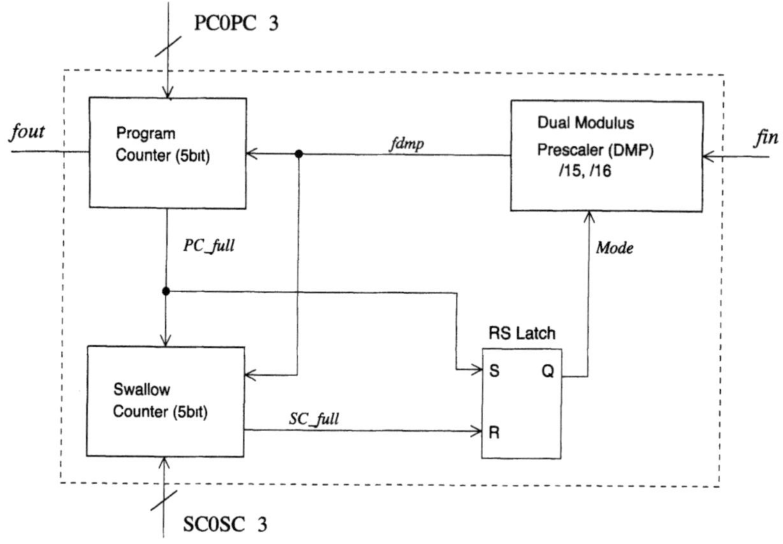

Figure 8.7.   System block diagram of the Integer-N divider.

- WCDMA division ratio = 369 = 24x16 - 15 = PxN-S → P=24 , S=15 and N=16 division ratio = 357 = 23x16 - 11 = PxN-S → P=23 , S=11 and N=16

The integer-N divider is also used in fractional-N operation. There is a control signal from the Delta-Sigma Modulator (DSM) to the Integer-N divider.

- when the control signal is low i.e., 0 , division ratio of Int-N is N which is defined by the program counter inputs (PC0-PC3) and the swallow counter inputs (SW0-SW3).

- when the control signal is high i.e, 1 , the division ratio of Int-N is (N+1).

Since we need increase the division ratio by 1, a control logic circuit is inserted between the swallow counter control inputs (SWC0-SWC3). This circuit simply has the following inputs; 4-bits division control for swallow counter from outside (SC0-SC3) 1 bit control signal (CS) from DSM and 4-bits division control output which are connected to (SWC0-SWC3). During the WCDMA mode, CS is set to zero, so SWC0=SC0, ..., SWC3=SC3. When the GSM mode is on, CS signal is changed between 1 and 0 to obtain division ratios N and (N+1) for fractional-N division operation.

4.   Implementation of the Integer-N Architecture

In this section, the implementation of blocks in Integer-N architecture is described. A system reset signal sets the RS latch and counters to the correct states at start-up.

DMP input, *fin*, comes from the VCO output buffer as shown in Figure 8.7. Mode input which set division ratio of DMP comes from the RS latch. The output of DMP drives the program and swallow counters. When the mode is set to 1 or 0 , DMP divides by 15 or 16, respectively.

The RS latch function is implemented by using an edge-triggered D flip-flop (DFF) with clear and set controls. Q output of the DFF is connected to *mode* input of the DMP. The swallow counter output, *SC_full* , is connected to R input of the RS latch. The program counter output, *PC_full* , is connected to S input of the RS latch. Also, the system reset signal is also connected to the RS latch clear input.

The swallow counter basically counts up to number (S) provided by control inputs (SC0-SC3), then it makes RESET output high to reset the RS latch. It is constructed by a 4-bit synchronous counter and XOR gates. The outputs of this counter are compared to SC0-SC3 inputs by simply using XNOR gates. When the counter is full, the RESET output activates the RS latch so that the RS latch output changes DMP mode. It also has clear input which is connected to the system reset signal.

The program counter counts up to number (P) provided by control inputs (PC0-PC3), then it resets the RS latch. It is constructed by a 5-bit synchronous counter and outputs of this counter are compared to SC0-SC3 inputs by simply using XNOR gates.

4-bit Synchronous Counter

A 4-bit synchronous counter with Gray like coding is designed to reduce delay and to implement race free logic. The Gray like coding of transitions of the counter states implies that only one output changes during the transition from one state to the other state. The state transition table and corresponding counting numbers are given in Table 8.5. Notice that the binary value does not necessarily correspond to actual number. When the counter outputs are compared with counter control inputs, the control inputs must be given in a binary number such that the binary input corresponds to the desired number. For example, for counting 12, the binary input = 1101. The 5-bit counter is implemented by adding 4-bit synchronous counter to a 1- bit asynchronous counter. The following are the equations for the 4-bit counter;

number	present state					next state			
#	$q3$	$q2$	$q1$	$q0$		$Q3$	$Q2$	$Q1$	$Q0$
1	0	0	0	0		0	0	0	1
2	0	0	0	1		0	0	1	1
3	0	0	1	1		0	0	1	0
4	0	0	1	0		0	1	1	0
5	0	1	1	0		0	1	0	0
6	0	1	0	0		0	1	0	1
7	0	1	0	1		0	1	1	1
8	0	1	1	1		1	1	1	1
9	1	1	1	1		1	1	1	0
10	1	1	1	0		1	1	0	0
11	1	1	0	0		1	1	0	1
12	1	1	0	1		1	0	0	1
13	1	0	0	1		1	0	1	1
14	1	0	1	1		1	0	1	0
15	1	0	1	0		1	0	0	0
16	1	0	0	0		0	0	0	0

Table 8.5.   State transition table for 4-bit counter.

$$D_0 = \bar{q}_1 q_0 + \bar{q}_1 q_2 + \bar{q}_1 \bar{q}_3 + \bar{q}_3 q_2 q_0$$

$$D_1 = q_1 q_0 + q_0 \bar{q}_3 + \bar{q}_3 \bar{q}_2 q_1 + q_3 \bar{q}_2 q_0$$

$$D_2 = \bar{q}_3 q_2 + q_1 q_2 + q_2 \bar{q}_0 + \bar{q}_3 \bar{q}_0 q_1$$

$$D_3 = q_3 q_2 + q_0 q_3 + q_1 q_3 + q_0 q_1 q_2$$

$$(8.2)$$

Interface Circuit for Fractional-N operation

The interface circuit simply changes the division ratio between N and (N+1) with control signal from the delta-sigma modulator (DSM) during the GSM

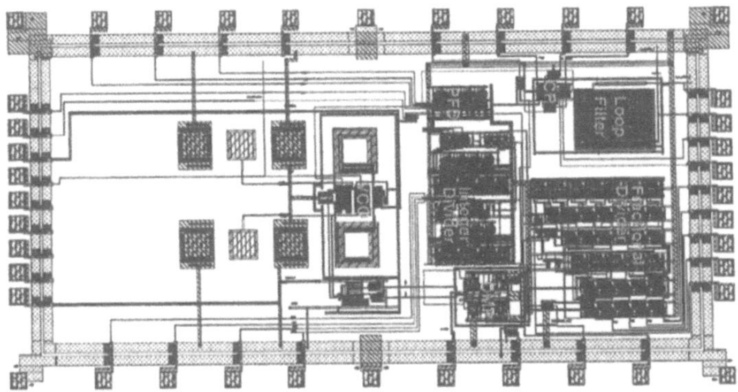

Figure 8.8. Layout view of the whole synthesizer.

mode. When the control signal is high, 1 , and low, 0 , the division ratio becomes (N) and (N+1), respectively. The logic circuit from the 4-bit counter is simply used for this operation. Integer-N block is used in both modes of operation while Interface and DSM blocks are only used during GSM operation. The hardware sharing of the integer-N block for both modes saves a large die area since it occupies most of the die area. 70% of the die area is shared among the two wireless standards.

Layout

The synthesizer is implemented in a 0.5μm epi-CMOS technology with 3 metal layers and 2 poly layers. The chip layout view is shown in Figure 8.8. Before submission of the chip, LVS is performed on each major block to verify connectivity. Each digital output is connected to bonding pads through a simple buffer (two inverter chain) to prevent loading the pertaining digital node with bonding pad capacitance. To measure the loop filter output voltage, a simple analog buffer is designed by using a simple source follower topology.

5.   Summary

A fully integrated dual-mode frequency synthesizer for GSM and WCDMA standards is presented. It is shown that the hardware sharing can be maximized by a careful frequency planing and architecture selection. A hybrid Integer-N and fractional-N architecture is proved to be useful for hardware sharing and performance improvement. A dual-band RF VCO is designed.

Chapter 9

PLLs FOR IEEE 802.11 a/b/g WLANs

This chapter discusses VCO and PLL design for a multi-band tri-mode fully integrated WLAN radio in $01.8\mu m$ CMOS. The VCO/PLL are discussed in the context of the radio architecture and frequency plan being used. Design of other PLL blocks such as prescaler, VCO buffers and loop filter are also discussed. Phase noise optimization for meeting the stringiest performance requirements of OFDM is described. Experimental results from test chips are given.

1.   Why Multi-band Tri-mode WLANs?

Local area networks (LANs) are being deployed extensively worldwide using a well-established copper-cable infrastructure. Wireless local area networks (WLANs) have become available as well as other emerging wireless applications in Instrumentation, Scientific and Medical (ISM) bands [74], including Bluetooth and HomeRF. The initial WLAN applications have used an unlicensed 2.4GHz ISM band. The 2.4GHz ISM band has been used for standards including cordless phones, Bluetooth, HomeRF and microwave oven in addition to WLAN. The 2.4GHz band is heavily occupied with other standards, which causes interference leading to slower data rate for WLAN. A 2.4GHz higher data rate (54 Mb/s) WLAN standard (802.11g) based on OFDM is proposed by IEEE to reduce multipath fading. 802.11g is also backward-compatible to the already deployed 802.11b (11 Mb/s) standard. An unused 5GHz band also exists, which is cleaner and has more bandwidth to accommodate higher data throughput (54 Mb/s or higher). There will be need to move to the 5GHz band (802.11a or Wi-Fi5) in the near future with increasing demand for higher data rate. Therefore, WLAN radios must be able to handle both of the above mentioned frequency bands for smooth transition between standards and increased user capacity by using both bands. A forecast for the WLAN market growth is shown in Figure 9.1([75]), where significant growth is anticipates for 802.11 a/b/g solutions.

In this context, a wide band radio transceiver capable of supporting different frequency bands and standards is a very desirable solution for WLAN applications. A careful frequency planning and architecture selection can simplify radio implementation by component sharing and higher integration level in different operation modes.

Fully integrated PLL frequency synthesizers, part of an IEEE 802.11a/b/g fully integrated transceiver [76–78], are implemented in a $0.18\mu m$ CMOS technology. Design of these PLL frequency synthesizers is presented in this chapter. First, frequency plan, radio architecture and synthesizer specifications are over-viewed. Next, phase noise optimization and trade-offs for designed synthesizers are presented. Loop filter design issues and solutions in an integrated transceiver environment are also investigated.Implementation of the RF and IF PLL frequency synthesizers are presented. Integration issues and block design challenges are identified and discussed, and also solutions are presented. Finally, measured results and comparison with published works are presented.

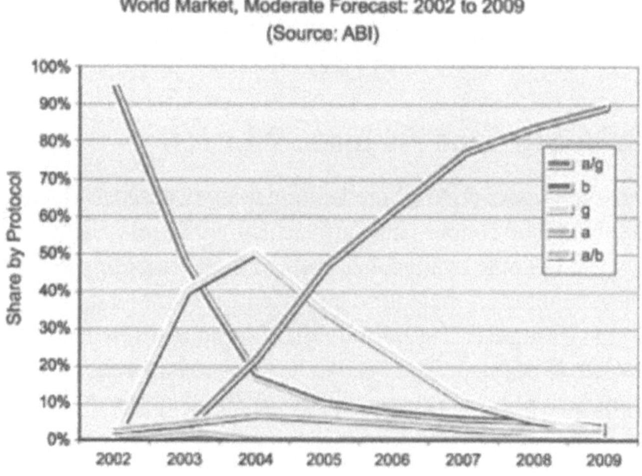

Figure 9.1.   Wireless LAN growth [75].

2.   Frequency Plan, Radio Architecture and PLL Specifications

The frequency plan is an important aspect of the implemented multi-band and multi-standard radio architecture. A great deal of hardware reuse leading to lower power consumption and smaller die area is achieved by careful frequency planning. Another important aspect of this architecture is that the frequency synthesizers are shared between the transmitter and the receiver, and hence simplifying the system complexity and implementation. This is possible because IEEE 802.11 a/b/g standards use time-division duplexing (TDD).

The frequency plan as shown in Figure 9.2 is constructed to cover the currently existing RF bands of the standards. The first local oscillator has two distinct frequencies, 3840MHz and 4320MHz, to translate channels from the RF bands to an IF frequency range between 1340MHz-1535MHz. The first RF LO frequency at 3840MHz translates channels from 2.4GHz RF band and 5.15-5.35GHz RF band to the IF band. The second RF LO frequency at 4320MHz translates channels from 5725-5825MHz RF band to the IF band. A single IF frequency would produce a wide IF band which would pose great challenges to IF frequency synthesizer design because of very large division ratios in phase-locked loops (PLLs) used to generate local oscillator frequencies. Therefore, the RF LO frequencies are chosen such that the IF span, in this case 1340MHz-1505MHz, is relatively small. This facilitates the operation of the circuits following the down conversion. From Figure 9.2, it can be seen that the distinct RF LO frequencies are chosen such that they are located approximately between the bands.

Wide-band IF Architecture

The radio architecture shown in Figure 9.3 is based on a wideband IF receiver and a two-step up conversion transmitter. The RF band, 5GHz or 2.4GHz, is first down converted to a common IF frequency using a fixed RF LO (LO1). The LO1 frequency is chosen to be halfway between the two RF bands. The IF signal (1300-1500MHz) is then down converted to baseband using a complex I/Q mixer. The channel is selected using a variable IF LO (LO2) for the IF to baseband down conversion.

The frequency plan adopted in this design achieves four basic functions: inherent image rejection, low phase noise, better IQ matching, and maximum hardware share. This leads to a low power, small die size multi-band multi-standard radio solution. Since the difference between the two frequency bands is more than 2GHz, no image rejection scheme, on or off chip, is required. More than adequate image rejection is achieved by the RF band select filter and tuned circuits in the LNA and mixer. The 5GHz band (almost 1GHz wide) is translated to a much narrower band in IF. This relaxes the design requirements of the LO2 circuitry, resulting in improved phase noise performance. LO1, although operating at a high frequency, is fixed. This allows a low division ratio in the LO1 PLL, giving lower phase noise. Moreover, I/Q modulation and demodulation is moved to the IF range which eases I/Q matching. This multi band/standard architecture achieves maximum hardware share resulting

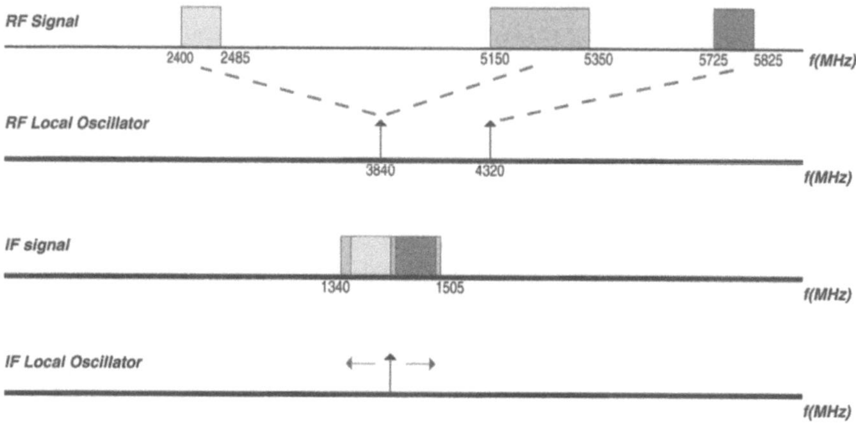

Figure 9.2.   Frequency plan for multi-band and multi-mode operation WLAN applications.

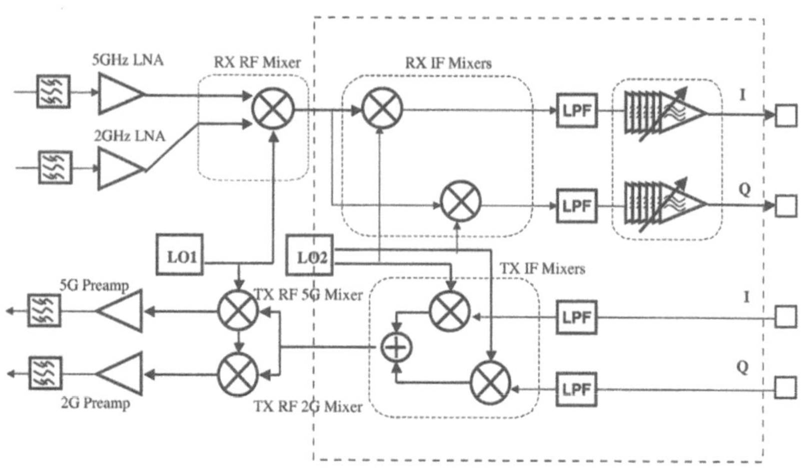

Figure 9.3.   Wideband-IF radio transceiver architecture.

Parameter	LO1	LO2
Integrated phase noise (dBc)	-33	-33
Settling time, ±20% (μsec)	224	224
Frequency resolution (MHz)	160	1
Reference frequency (MHz)	20	1
VCO PN at 1MHz(dBc/Hz)	-110	-110
VCO tuning range	3.6-4.4GHz (20%)	1.3-1.5GHz (21%)
Reference spur (dBc)	-48	-48
Harmonic (3rd) (dBc)	-30	-40
IQ phase mismatch	-	$< 1^\circ$
IQ amp. mismatch(dB)	-	±1

Table 9.1. Summary of the RF (LO1) and IF (LO2) PLL specifications.

in considerable saving in chip area. The LNAs, TX RF mixers and preamplifiers are the only parts that are not shared. The architecture supports the entire 5GHz band (4.9-5.9GHz). However, the implementation reported here is optimized for the lower 5GHz band(5.15-5.35GHz).

Two PLL frequency synthesizers are designed and implemented for the required LO1 and LO2 signals. An integer-N PLL architecture is employed in both synthesizers. The LO1 synthesizes the fixed RF LO (3840MHz) from a 20MHz reference signal. The LO2 synthesizes the variable IF LO with a 1MHz step in the frequency range of 1300MHz to 1500MHz. An on-chip crystal oscillator generates the 20MHz reference signal. Summary of the specifications for LO1 and LO2 synthesizer are listed in Table 9.1.

Zero-IF Architecture

The designed frequency synthesizers, LO1 and LO2, can easily be used to generate quadrature LO signals for multi-band zero-IF architecture (or low-IF) as shown in Figure 9.4. The frequency components LO1+LO2 and LO1-LO2, generated by mixers M3 and M4 in the synthesizer block, are used to drive mixers M1 and M2 in the receiver.

Implementations of the LO1 and LO2 are discussed next for the wide-band IF architecture shown in Figure 9.3.

3.   Phase Noise Optimization and Trade-offs

The SNR (or the BER performances) that can be achieved in an OFDM system (IEEE 802.11a) with non-ideal noisy local oscillator can be approximated by integrating its phase noise power density over the channel bandwidth. The integrated phase noise specification for IEEE 802.11a (also 802.11g) is defined

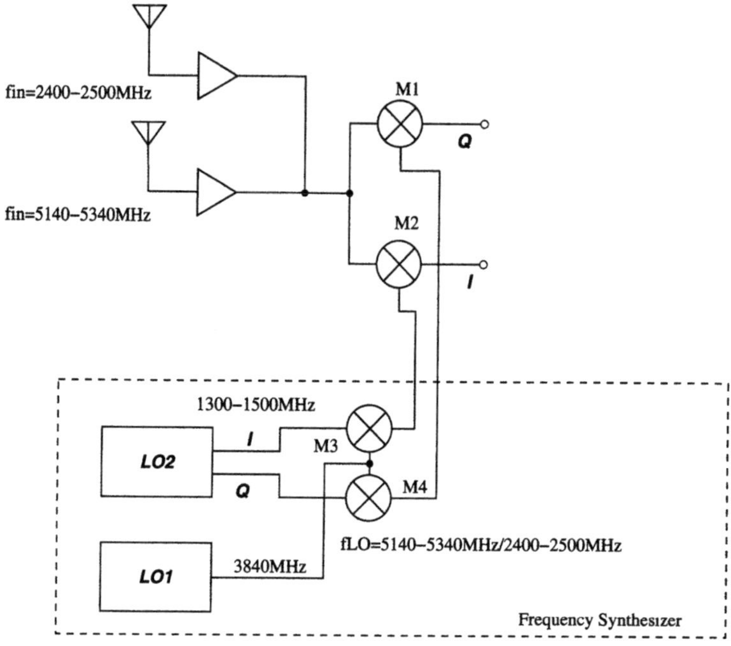

Figure 9.4. Quadrature LO generation in a zero-IF (or low-IF) receiver using LO1 and LO2. LO1 is fixed while LO2 is used for frequency synthesis (or vice versa).

as the total double sided integrated noise from $f_1 = 30KHz$ to $f_2 = 8MHz$ which must be $-30dBc$ or better [79]. For a given single-sideband (SSB) phase noise profile, $L(f)$, of a local oscillator, the integrated phase noise is defined as,

$$L_{tot} = 2 \int_{f_1}^{f_2} L(f) df \qquad (9.1)$$

where factor 2 is used to account for double-side band (DSB) noise. We are interested in the answers to the questions "How should phase noise be characterized?" and "What phase noise is really necessary?" in order to properly specify the phase noise of local oscillators to meet system requirements without over-specifying. A typical SSB phase noise profile of PLL frequency synthesizer (local oscillator) is shown in Figure 9.5. L_o is the PLL close in phase noise. B is the PLL loop bandwidth. The PLL noise profile in Figure 9.5 is also the VCO noise profile for offset frequencies $f > B$. The SSB VCO phase noise (for $f > B$) in terms of PLL loop bandwidth and close-in noise can be expressed as;

$$L_{vco}(f) = \frac{L_o}{(f/B)^2} \qquad (9.2)$$

Using Equations 9.1, 9.2, and Figure 9.5, we can derive the relationships among the PLL bandwidth, the PLL close-in noise value (L_o), and the VCO phase noise at a specific offset frequency from a given integrated noise value. That is,

$$L_{tot} = 2 \int_{f_1}^{f_2} L(f)df = 2 \int_{f_1}^{B} L_o df + 2 \int_{B}^{f_2} \frac{L_o}{(f/B)^2} df \qquad (9.3)$$

This leads to,

$$L_{tot} = 2L_o(B - f_1) + 2B^2 L_o \left(\frac{1}{B} - \frac{1}{f_2} \right) \qquad (9.4)$$

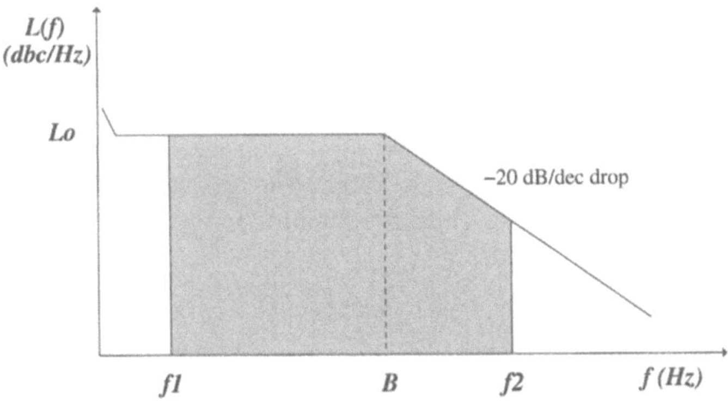

Figure 9.5.   A typical PLL SSB phase noise profile.

The PLL bandwidth, B, is the key parameter in determining PLL close-in noise and VCO noise at large offset frequencies (for $f > B$) for a given total integrated noise value of $L_{tot}(dBc)$. For the integrated noise value of $L_{tot} = -33dBc$, the required PLL close-in noise value, L_o, and the VCO phase noise value at an offset frequency of $1MHz$ are plotted for different practical values of the PLL loop bandwidth, B as shown in Figure 9.6. The lock time of the PLL is also taken into consideration when determining the value of B since the lock time and the loop bandwidth are inversely proportional to one another. Figure 9.6 suggests a trade-off between the VCO noise and the PLL-close in noise for a given B. The PLL close-in noise is related to the feedback divide ratio, N and the reference frequency values as suggested by Equation 3.26. Equation 3.26 is re-written here.

$$L_o(dBc/Hz) = L_{pll,nf} + 20\log(N) + 10\log(f_{ref}) \qquad (9.5)$$

If we use the same CP and PFD circuits, we will get different close in noise values for LO1 and LO2 PLLs. LO1 PLL uses a reference frequency of 20MHz,

B (kHz)	Lo (dBc/Hz)	$Lvco@1MHz$ (dBc/Hz)	$L_{nf,LO1}$ (dBc/Hz)	$L_{nf,LO2}$ (dBc/Hz)	L_{tot} (dBc/Hz)
25	-81.4	-113.4	-200.4	-204.4	-33
50	-85.3	-111.3	-204.0	-208.2	-33
100	-88.6	-108.7	-207.3	-211.5	-33
200	-91.8	-105.8	-210.5	-214.7	-33
400	-94.8	-102.8	-213.5	-217.8	-33

Table 9.2. PLL close-in noise, VCO noise at 1MHz offset, and PLL noise floor for different values of loop bandwidth.

and divide ratio is 200 while LO2 PLL uses a reference frequency of 1MHz, and divide ratio is in the range of 1300-1500. In order to optimize the PLL close-in noise to meet required total integrated phase noise specification, the following methods are employed;

- Off-chip loop filter is used for both PLLs to optimize the loop bandwidth, B. The reasons to use off-chip loop filter are further explained in Section 4 dealing with loop filter design.

- The CP output current is made programmable for both PLLs. 2-bit control word selects one of 4 different CP current values; 100uA, 200uA, 500uA, and 1mA. Increasing CP current reduces CP noise contributions, and hence it can be used to compensate divide ratio and reference frequency change.

- PLL noise is modeled by using PLL noise model presented in Chapter 3. All noise sources of PLL blocks are extracted through measurement and simulation.

Table 9.2 lists phase noise requirements for different PLL loop bandwidths. LO2 PLL requires lower phase noise floor due to higher divide ratio. This is achieved by using larger CP current for LO2.

4.   Loop Filter Design

Loop filter design is an important aspect of the frequency synthesizer design since most performance parameters (phase noise, stability, lock time, reference spurs) of the synthesizer depend on the loop filter parameters. A major design consideration in the design of loop filter the PLL is to minimize the integrated phase noise to meet the specification by changing charge pump current and loop bandwidth while meeting the lock time specification and reference spur levels. For optimum design of the loop filter, one question may arise; "should

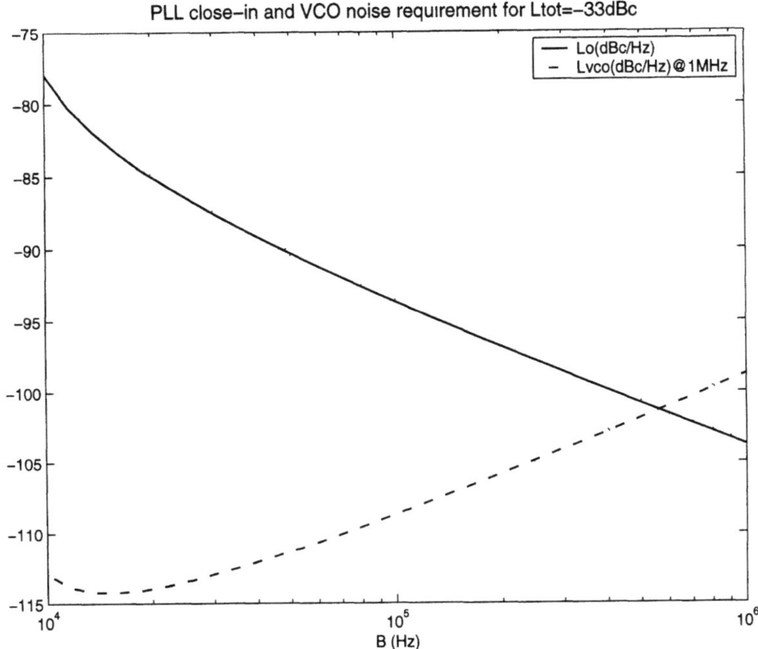

Figure 9.6. Phase noise requirements for VCO and PLL close-in noises for a constant integrated noise value of $L_{tot} = -33dBc$ for different values of B.

loop filter be placed off-chip or on chip, and why?". The loop filter parameters (R_Z, C_z, and C_p) are designed for given PLL parameters; I_{CP}, K_{VCO}, and N and for specified loop bandwidth (ω_c) and phase margin (ϕ_m) (Chapter 2). The PLL parameters; VCO gain and feedback divide ratio can be considered as design constants; however we have freedom to chose charge pump current value. Each charge pump current value will result in a set of loop filter parameters for given phase margin and loop bandwidth values. Smaller charge pump current results in smaller loop filter capacitance values (Equation 2.11); and hence eases the loop filter integration. Larger charge pump current reduces the close in phase noise (Equation 3.18); and hence it reduces integrated phase noise. Selection of a charge pump current value depends on the PLL requirements. If the integrated phase noise requirement is relaxed or not constrained, this translates into relaxed close-in PLL noise. When the close-in noise is relaxed or not constrained, the loop filter can be integrated (on-chip) by making CP current as small as possible in order to make loop filter capacitor values small enough so that their physical sizes are practically implementable. If the integrated phase noise is tightly specified, then the CP current value and loop bandwidth are concurrently changed to meet the integrated noise specification. Therefore, the loop filter must be placed off-chip for noise optimization purposes; and this is the case for this design.

Placement of PLL loop filter off-chip in a fully integrated environment presents implementation challenges. A conventional loop filter architecture is shown in Figure 9.7(a). An additional pole is added to loop filter ($\omega_p = 1/R_3C_3$) in order to reduce phase noise contributions of the CP, reference spur level, and on-chip noise coupling. Implementation of the conventional loop filter architecture in Figure 9.7(a) in a fully integrated transceiver environment faces the following issues;

- The actual control voltage seen by the VCO, $VCNT$, is not the same as the loop filter output DC voltage, $Vloop\_filter$ due to different ground references. The loop filter DC voltage is referenced to PCB board ground while the VCO control voltage is referenced to on-chip VCO ground. The control voltage, $VCNT$, seen by VCO is equal to,

$$V_{cnt} = V_{loop\_filter} - I_{DC,osc}R_{gndosc} \qquad (9.6)$$

  Due to finite resistance of interconnect lines on-chip, ground line exhibits a resistance in the range, $R_{gndosc} = 5\Omega - 15\Omega$. VCO and the other PLL analog blocks draw a DC current of about $10mA$ in this design. This translates to a voltage difference of $50mV$-$150mV$ between loop filter and VCO control voltage. This further limits the VCO tuning range. In addition to DC voltage difference, any noise coupled to VCO ground line including ground bounces will reflect onto the control voltage. These are then converted into phase noise and spurious sidebands at the VCO output. Ground bounce is especially important for a fully integrated transceiver. Ground Bounce is a voltage oscillation between the ground pin in a component package and the ground reference level on the die (on-chip). It is caused by a current surge passing through the lead inductance of the package. Even though VCO ground is relatively quite, the mutual coupling between bond wires carries ground bounces from other ground lines to the VCO ground line.

- Noise and on-chip signals coupled to the interconnect line between CP output and the loop filter via the substrate, other blocks, and interconnects will appear on the control voltage. These coupled noise and on-chip signals on the control voltage translates into phase noise and spurious signals at the VCO output through an FM effect.

- Similarly, noise and on-chip signals couples into the interconnect line between loop filter output and VCO control voltage input from the substrate, other blocks, and interconnects since the control voltage is a single ended signal. These signals translates into phase noise and spurious signals at the VCO output through an FM effect.

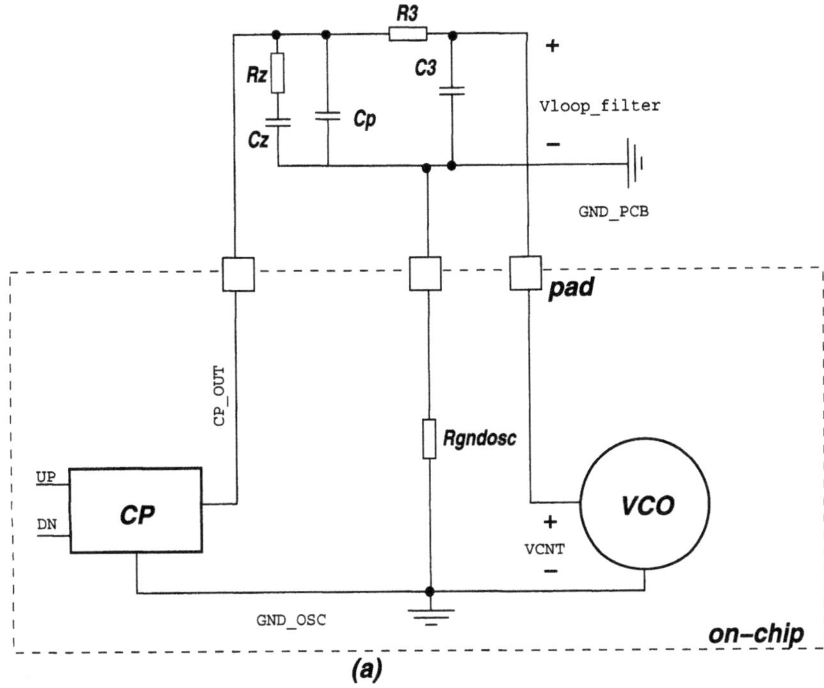

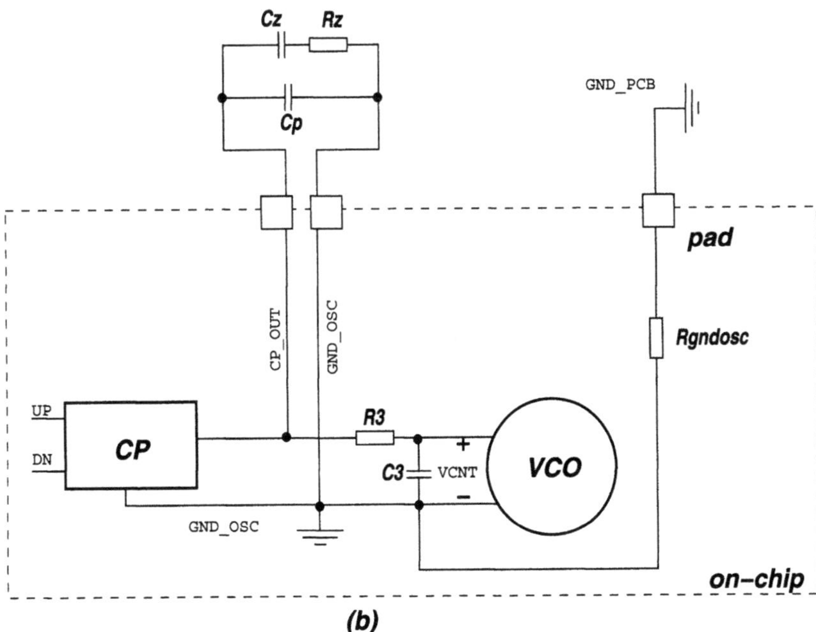

Figure 9.7. Loop filter architecture (a) conventional loop filter architecture (b) proposed loop filter architecture.

Noise issues with the conventional loop filter architecture can be solved by using fully differential CP and loop filter at the expense of die area, power consumption and complexity. However, DC voltage difference due different ground references is still an issue with differential implementation.

A loop filter architecture is proposed to solve the issues discussed above with the conventional architecture. The proposed loop filter architecture is shown in Figure 9.7(b). Ground line for the loop filter is taken from on-chip ground as a separate signal line. This leads to a loop filter voltage equal to VCO control voltage, and hence avoiding DC voltage drop between loop filter and VCO control voltage. The noise coupling to the control voltage on the interconnect lines between loop filter and VCO is suppressed since the control voltage and ground line behave as a differential line and reject common mode noise signals. For further noise suppression, the additional pole components are placed in close proximity to the VCO on chip. This loop filter architecture removes DC voltage drop due to ground reference difference and provides immunity to on-chip noise coupling through common mode rejection. It also suppresses ground bounce.

5.   The RF (LO1) Synthesizer Implementation

The RF (LO1) frequency synthesizer is designed for RF LO1 generation. The distinct RF LO1 frequencies at 3840MHz and 4320MHz are used to convert the RF signals from 2.4GHz and 5GHz bands to intermediate frequency(IF) band around 1400MHz. The top level architecture of LO1 synthesizer is shown in Figure 9.8.

The LO1 synthesizer is based on an integer-N phase-locked loop architecture. The feedback divider is formed by a fixed prescaler with a divide ratio, P=8, and programmable low frequency divider with divide ratios 24,26, and 27. Use of a fixed prescaler at 4GHZ saves power consumption. Due to the adopted frequency plan, a dual-modulus prescaler (DMP) was not necessary to synthesize the required frequencies, instead programming of the division ratio is moved to a low frequency range ($< 600MHz$).

Critical building blocks in the LO1 architecture are 4GHz VCO (wide tuning range, low VCO gain, phase noise), VCO buffer, prescaler operating at 4GHz range, and loop filter design with optimum phase noise. Design issues and solutions for these building blocks are discussed next.

4GHz RF VCO Design

LO1 VCO requires a wide band operation with a 20% tuning range (3.6GHz-4.4GHz). Fortunately, the PLL needs to synthesize only discrete frequencies far apart in the frequency spectrum, i.e., VCO tuning range does not have to cover any band continuously. This allows us to divide the VCO tuning range

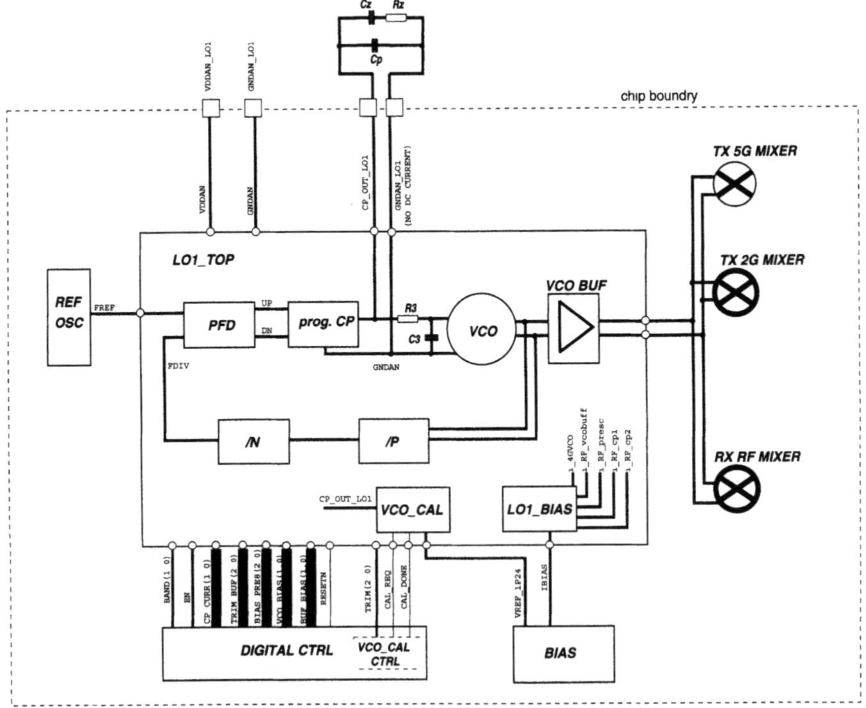

Figure 9.8. The RF (LO1) Frequency synthesizer architecture.

into small bands; and hence smaller VCO gain can be obtained. A developed digital calibration circuit selects the optimum sub-tuning band for the desired frequency. Low VCO gain helps in lowering the VCO phase noise. Design of calibration circuit is presented in Chapter 6.

The VCO exclusively uses PMOS devices to take advantage of lower flicker noise densities compared to NMOS devices. The simplified schematic of VCO is shown in Figure 9.9. The tank consists of a center-tapped (differential) inductor, accumulation mode NMOS varactor diodes and switched capacitors. Critical performance parameters in this VCO design are phase noise, flicker noise corner, supply sensitivity, bias current sensitivity, harmonic levels, and power consumption. Succesful design of RF VCOs in CMOS technology require accurate simulation of performance parameters and good device models (RF MOS, varactor, and inductor models). EldoRF [23] is used for RF simulations. ASITIC [64] is used for inductor modeling. Test structures for varactors and inductors are fabricated. These test structure are characterized with on-wafer S-parameters measurements (Chapter 10).

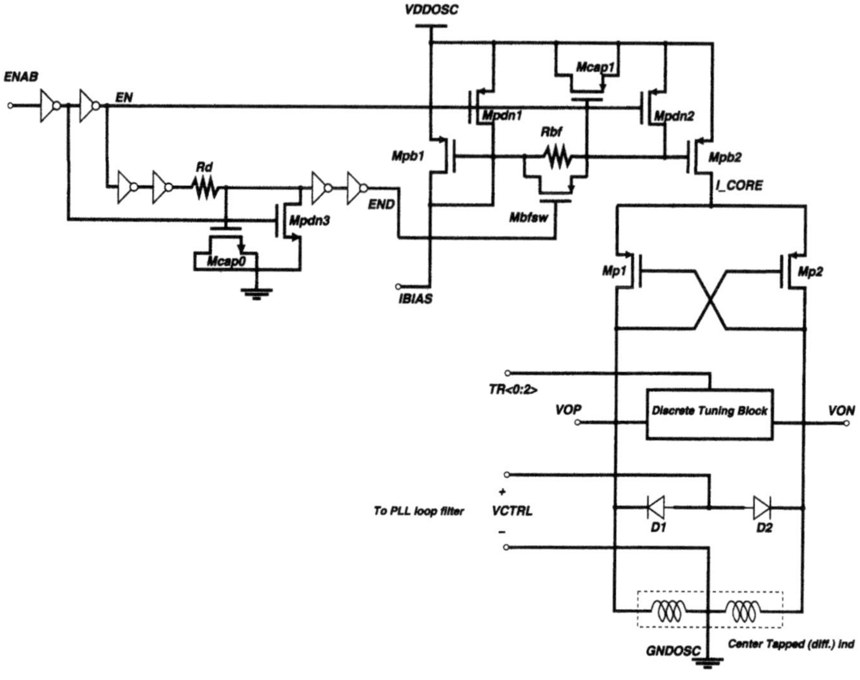

Figure 9.9.   Simplified schematic of 4GHz VCO.

A dynamic bias filter is developed for filtering reference bias noise and for suppressing bias sensitivity. Reference bias current is taken to LO1 from a on-chip band-gap circuit. This reference current is distributed for LO1 blocks. The reference current noise is the sum of band-gap and bias mirrors noise. This noise is converted to phase noise in the VCO. Another purpose of the bias filter is to suppress bias sensitivity of VCO. The reference bias current from band-gap is disturbed during RX to TX or TX to RX switching. A dynamic transient disturbance occurs in the band-gap generated bias current during the switching since a large amount of DC current is switched (in the order of 100mA). This dynamic variation in the reference current, in turn, results in disturbance in VCO output frequency and can pull-out the VCO from lock. Bias low-pass filter is formed by a resistor, Rbf, and MOS capacitance, $Mcap1$. The filter corner frequency is set to $2.5KHz$. The corner frequency is chosen much smaller than PLL loop bandwidth so that PLL is able to track VCO frequency variation due to any disturbance on the bias current. VCO pulling in the integrated environment is discussed in Chapter 6. Noise from reference current is also suppressed below the lower phase noise integration limit frequency, $30KHz$. However, the small low-pass frequency result in a large time constant which in turn leads to a long start-up time for the VCO. A long start-up time of VCO may prohibit

Parameter	min	typ	max	unit
Output frequency	3600		4400	MHz
Phase Noise at 1MHz		-120	-117.4	dBc/Hz
Supply sensitivity			4	MHz/V
Bias sensitivity			0.75	MHz/uA
Load sensitivity			0.66	MHz/fF
Start-up time			5	usec
Gain	70	120	180	MHz/V
2nd HM			-53	dBc
3nd HM			-30	dBc
Current	5	5.7	6.3	mA

Table 9.3. Simulated characteristics of LO1 VCO.

PLL from meeting lock time requirement. A dynamic power up/down scheme is added to the bias filter of the VCO to speed up the VCO in power up and down modes. A switch transistor, $Mbfsw$, is placed in parallel to bias filter resistor, Rbf. The switch transistor by-passes the bias filter resistor, Rbf, at power up, and hence the bias filter time constant is lowered. Lowered filter time constant speeds up the start-up of VCO. The switch transistor, $Mbfsw$, is kept ON during the short (about $5\mu sec$) time by END a delayed version of the power-up signal, $ENAB$. The delayed version of power-up signal, $ENAB$, is generated by inverters and low-pass filter formed by Rd and $Mcap0$. The power down device, $Mpdn3$, is used to speed up the power down of the VCO.

Table 9.3 lists the simulated characteristics of 4G VCO. Typical (typ) values correspond to simulated results for room temperatures, and nominal bias values with typical process corner models. Maximum (max) values correspond to -30 C degrees, fast process corner models, and 10% increased nominal bias values. Minimum (min) values correspond to 80 C degrees, slow process corner models, and 10% decreased nominal bias values.

VCO Buffer Design

LO1 VCO output is to drive three RF mixers; RX RF mixer, TX 2GHz RF mixer, and TX 5GHz RF mixer. It is possible to connect RF mixers' LO inputs together since one of the mixers is active at a given time. An equivalent circuit model for a mixer differential LO input is shown in Figure 9.10. A single mixer LO input parameters are $C_{pd} = 80fF$ and $R_{pd} = 12k\Omega$. RF mixers' LO input differential impedance is quite high compared to VCO tank impedance. VCO equivalent parallel resistance is about $R_p = 400\Omega$ at 4GHz. Parallel connected mixers' LO inputs will present an equivalent load of $240fF$ and $4k\Omega$ to VCO resonator if they are directly connected to VCO output. VCO is able to drive mixers directly; however, this is avoided for two major reasons;

- Mixers are placed at far apart at top layout. Long signal routing lines from VCO output to mixers' LO input will add a large amount of capacitance and inject noise from all over the chip to VCO resonator.

- The turn around time between RX and TX operation is $8\mu sec$ for IEEE 802.11a/g. During turn around, the used RF mixer is enabled and the other mixers are powered down to save power. Switching ON and OFF mixers will create change in the load seen by the VCO tank, and this load change will pull out the VCO from lock. Pulling back the VCO to lock by the PLL takes longer time than the required turn around time.

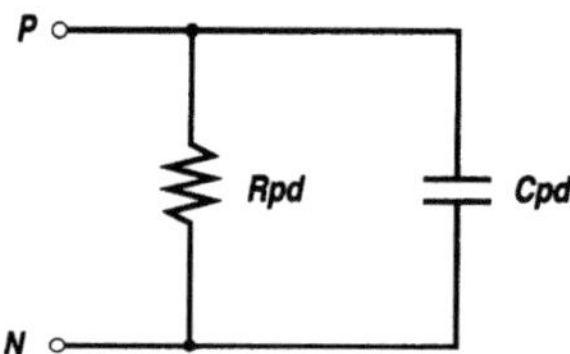

Figure 9.10.   Equivalent circuit model used to represent mixers' LO input.

A VCO buffer circuit is designed to isolate the VCO from on chip noise, capacitive loading, and pulling. The simplified buffer schematic is shown in Figure 9.11. Cascode topology is used for its high output impedance, and better reverse isolation. Buffer input is DC decoupled from VCO output by using C_{inn} and C_{inp}. Input devices, M_1 and M_2, are biased with internally generated bias voltage V_{bias1}. Inductor load is used for tuning out the capacitance seen at the buffer output and for increasing voltage headroom. Buffer output sees a large amount of capacitance from 3 mixers and routing lines. A resistive load, R_{load}, connected at the differential output is used to increase the bandwidth of the buffer. A trimming capacitance with 3-bit control word is used to extend the bandwidth of buffer. The trimming capacitance block is shown in Figure 9.12. Capacitance values are binary weighted to realize equal step sizes. The buffer needs to cover three discrete frequencies in the frequency range of 3.6-4GHz. A small valued resistor, R_{vdd} is added in series to the supply voltage to prevent the voltage value from exceeding technology specified value. A MOS device is permanently damaged if the voltage applied to its terminal exceeds a technology specified value.

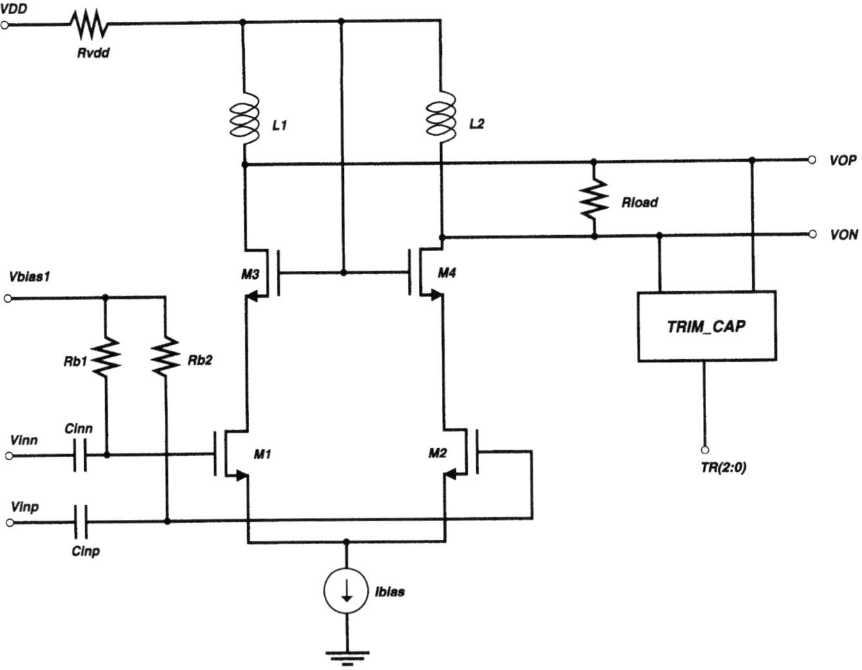

Figure 9.11. Simplified VCO buffer schematic.

One of the design challenges in the buffer design is the modeling of the differential interconnect lines. Differential RF signal line length between buffer output and mixers' LO input is in the order of mm. Long differential line causes RF signal loss (due to resistive and lossy substrate), noise coupling from substrate and the other blocks, capacitive loading. As a first order approximation, layout extraction tool is used to extract the capacitances between lines and substrate. This is sufficiently accurate for short lines ($< 10\mu m$). However, high frequency effects must be taken into consideration for longer lines in order to obtain accurate simulation results. A scalable model is developed for long differential RF line simulation. The cross-section of the differential line is shown in Figure 9.13(a). The line width, W, and spacing between lines, S are kept constant. The line length, L, is variable for the model in the simulation. The developed SPICE model is shown in Figure 9.13(b). ASITIC [64] is used to develop the SPICE model. The selection of the line width, W, and spacing S, involves trade-offs. Wider line width reduces resistive loss, but increases capacitive loading. Smaller spacing between lines increases common mode noise rejection from substrate and the other blocks, but also increases capacitive loading. For buffer design, the line width and spacing are chosen as $5\mu m$ for 3-5GHz RF signal frequency range.

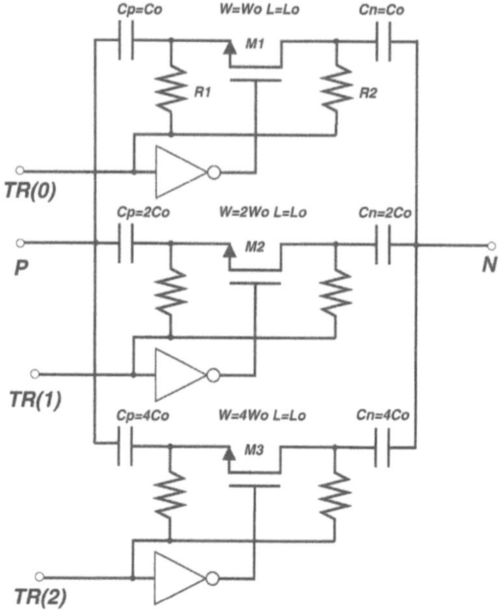

Figure 9.12.   VCO buffer trimming capacitance schematic.

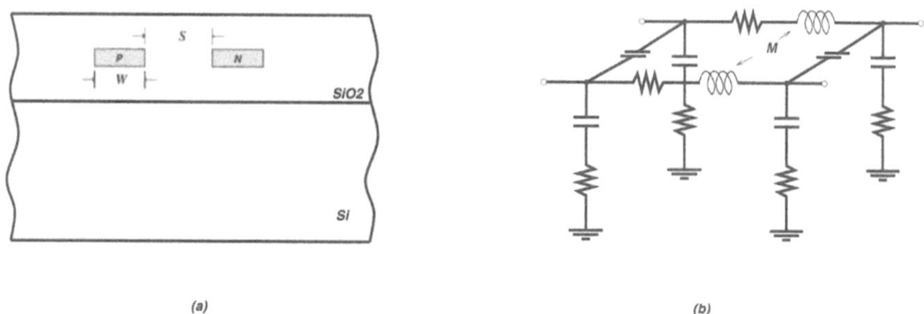

Figure 9.13.   (a) differential line structure (b) simulation model.

RF Prescaler Design

A frequency divider (prescaler) with fixed divide ratio of 8 operating at 4GHz is designed in order to lower the output frequency of VCO to level appropriate for the programmable counters in the LO1 PLL frequency synthesizer. The simplified prescaler architecture is shown in Figure 9.14. The prescaler is formed by 3 main blocks; high-speed divide-by-2 block, medium speed divide-by-4 block, and programmable bias block.

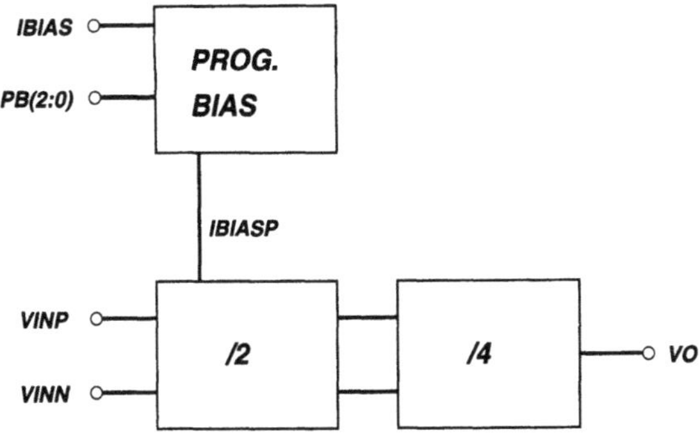

Figure 9.14. RF prescaler architecture.

The high speed divide-by-2 block is shown in Figure 9.15(a). The master and slave latches in Figure 9.15(a) are clocked by the differential VCO output. The CML type latch is used for high speed operation and low power consumption. The latch schematic is shown in Figure 9.15(b). When the input clock signal is high, V(CLK,CLKB)> 0, M5 is ON and M6 is OFF; and hence the latch circuit behave as sensing differential amplifier with M1, M2, M3, and M4 devices. PMOS transistors, M3 and M4, operating in triode act as load resistor for output. When the input clock signal is low, V(CLK,CLKB)< 0, M6 is ON and M5 is OFF; and hence the latch circuit behave as an differential pair with positive feed back; M7 and M8. The positive feedback latch the output signals (Q,QB) at their current level. A critical aspect in this implementation is the size of the PMOS transistors. The PMOS load transistors must be in triode while presenting as little parasitics as possible. Thus, very small PMOS transistor widths were used with minimum channel lengths. The DC biasing of input devices, M5, and M6, is set by $Vbias$ to optimize the performance of prescaler for high frequency operation and power dissipation. $Vbias$ is set by programmable bias current. The speed of the circuit is determined by the bias current of M5 and M6 together with the load capacitance seen at the output nodes, Q and QB. A higher current value can charge the load capacitor quickly; and hence it allows higher frequency operation. The load devices, M3 and M4, and input devices bias currents are optimized for operation in the frequency range of 3-5GHz. .Input devices' DC voltage, $Vbias$, is made programmable by using programmable bias current $IBIASP$ flowing through diode connected MOS device, M1 in Figure 9.15(a).

The medium speed divide-by-4 block is shown in Figure 9.16(a). It is formed by medium speed latches as shown in Figure 9.16(b). The latch consists of two sense devices (M1 and M2), a regenerative loop (M3 and M4), and two pull-up devices (M5 and M6) [80]. When CLK is high, M5 and M6 are off, and the latch is in the sense mode. When CLK is low, M5 and M6 are on, and the latch

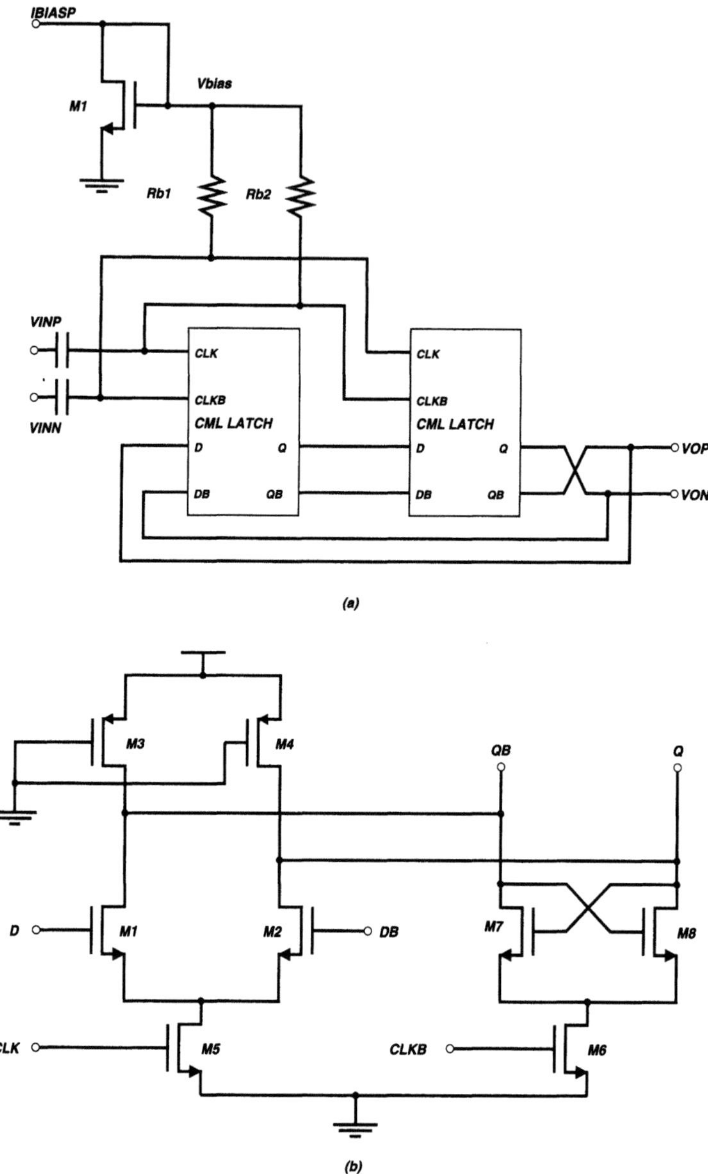

Figure 9.15.   (a) divide-by-2 schematic (b) CML latch.

is in the store mode [80]. The medium speed latch is chosen for low power consumption and small load capacitance to the divide-by-2 circuit.

The final prescaler circuit including programmable bias and all dividers draws less than 1mA current from 1.8V supply voltage. It can operate from 1-6GHz

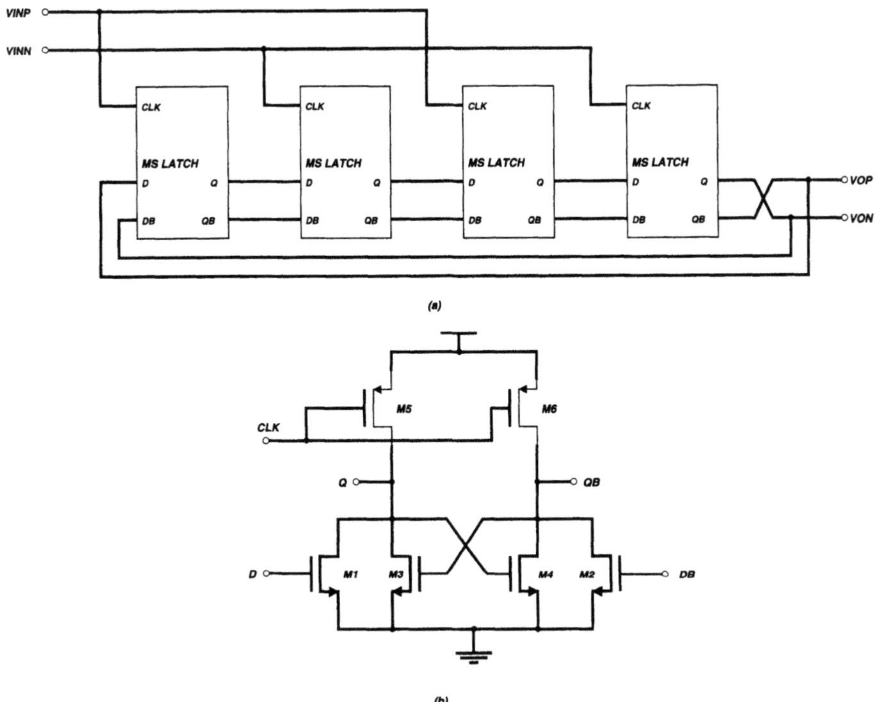

Figure 9.16.   (a) divide-by-4 schematic (b) medium speed latch.

frequency range with input differential amplitude 300mVpp or greater. The simulated phase noise of the prescaler is below other close-in noise contributor levels; and hence it does not degrade PLL noise.

6.   The IF (LO2) Synthesizer Implementation

The IF (LO2) frequency synthesizer is designed for IF quadrature LO signal generation in the frequency range of 1330MHz to 1480MHz with 1MHz steps. The top level architecture of LO2 synthesizer is shown in Figure 9.17. The LO2 synthesizer is based on an integer-N phase-locked loop architecture. The feedback divider is formed by a "pulse-swallow" divider architecture. The divider consists of a "dual-modulus prescaler" (DMP) with divide ratios , 32/33, a "program counter" (P-counter), and a "swallow counter" (S-counter). The VCO output frequency is set by controlling the P-counter and S-counter; and it is calculated by;

$$f_{out} = f_{ref}(PN + S) \tag{9.7}$$

where $N = 32$ and $f_{ref} = 1MHz$.

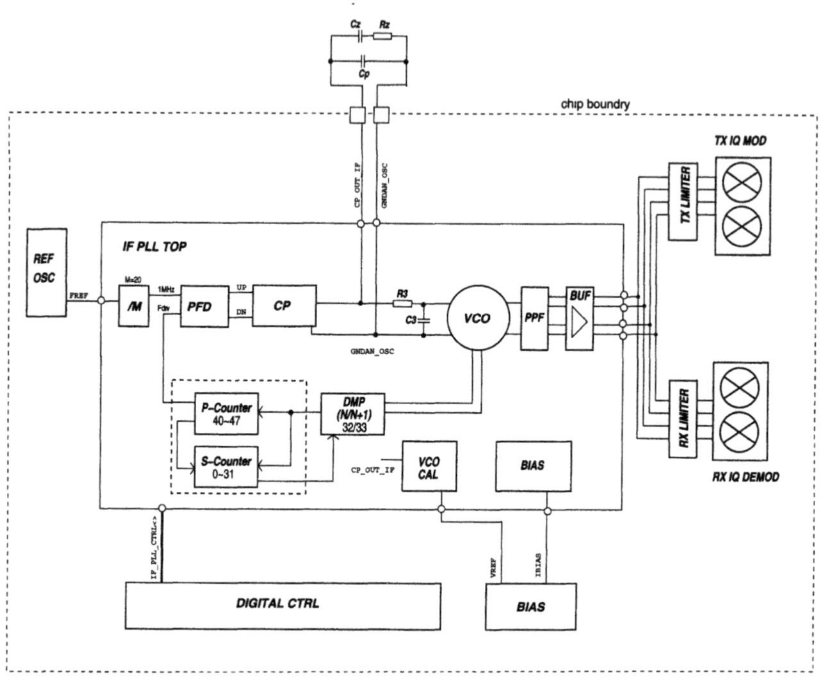

Figure 9.17.   The IF (LO2) Frequency synthesizer architecture.

Critical building blocks in the LO2 architecture are IF VCO (wide tuning range, low VCO gain, phase noise, low harmonic levels), IQ LO signal generation, dual-modulus prescaler design and VCO interface, and loop filter design with optimum phase noise. Design issues and solutions for these building blocks are discussed next.

VCO Design and Quadrature Signal Generation

LO2 VCO requires to cover continuously 1.3-1.5GHz tuning range to synthesize channel frequencies at 1MHZ steps. The required channels fall into the frequency range of 1330-1480MHZ. This continuous tuning range is divided into two sub bands to lower the VCO gain. The subbands are placed as 1330-1400MHz and 1401-1480MHz. The available CP output voltage (VCO control voltage) is in the range of 0.4-1.3V for high performance operation. This leads to an average VCO gain of 90 MHz/V. VCO tuning range supports 1250-1600MHZ frequency range to account for process variations. The entire tuning range is divided into 16 subbands. A digital calibration circuit similar to the one used in LO1 VCO selects the optimum sub-tuning tuning band from the 16 subbands. The same VCO topology as in LO1 VCO (Figure 9.9) is used

Parameter	min	typ	max	unit
Output frequency	1150		1680	MHz
Phase Noise at 1MHz		-123	-118	dBc/Hz
Supply sensitivity			1	MHz/V
Bias sensitivity			0.3	MHz/uA
Load sensitivity			0.15	MHz/fF
Start-up time			5	usec
Gain	92	124	167	MHz/V
2nd HM			-62	dBc
3nd HM	-53	-49	-39	dBc
Current	3.3	4.2	5.16	mA

Table 9.4. Simulated characteristics of the IF (LO2) VCO.

for the IF (LO2) VCO. The resonator and the active circuit are optimized for IF frequency range. Table 9.4 lists the simulated characteristics of the IF VCO.

Another critical design consideration in this VCO is to generate quadrature signals. One solution was to operate VCO at twice or 4 times higher frequency and use divide-by-2 or divide-by-4 circuits. These solutions has major drawbacks; (i) VCO output frequency gets close to the PA output frequency in both cases; this will cause VCO pulling by PA and addition VCO noise due to PA output leakage. (ii) Dual-modulus prescaler design is moved to higher frequency; this translates into implementation difficulties and high power consumption. A polyphase filter (PPF) is used for quadrature signal generation.

A two-stage polyphase filter (PPF) is used for quadrature LO signal generation. The schematic of PPF is shown in Figure 9.18. PPF is directly driven by VCO. The polyphase filter passes harmonics with an entirely different gain and phase than the fundamental, which upsets the duty cycle of the output waveform. This causes IQ phase and amplitude mismatch in the quadrature outputs. Amplitude mismatch is corrected by a limiter circuit. The IQ phase mismatch of LO2 signal is targeted for 1 degree or less. An IQ phase error margin of 0.5 degree is allocated for buffer/limiter and layout mismatches; and the maximum error margin for VCO and PPF is 0.5 degrees. The error from VCO and PPF depends on the relative harmonic amplitudes of the VCO. PPF input signal harmonics effect on IQ phase error is simulated for 2nd and 3rd harmonics. Figure 9.19 shows relative 2nd and 3rd harmonic levels versus IQ phase error. The 3rd harmonic level must be kept below -40dBc for 0.5 degree error. The sizes of RC components in PPF are determined by considering following requirements;

- The pole locations $1/R_1C_1$ and $1/R_2C_2$ are chosen to cover the IF bandwidth in the frequency range of 1.3-1.5GHz.

- The matching between the resistors and between the capacitors are simulated for IQ phase error contribution. Large components result in better matching of two devices. The minimum matching requirement leads to smallest value and the physical area of the components.

- Large resistor values will result in larger input impedance; and hence this results in less loading of VCO resonator tank. The resistor values are set at the largest possible values while maintaining the matching requirement for capacitors.

The output signal from the polyphase filter is first amplified with a two stage differential amplifier and then routed to the TX and RX limiter buffers. Separate limiter buffers were designed for the RX and TX in order to optimize the performance for each case.This way the output amplitude of the buffers remains nearly constant over a wide input amplitude range.

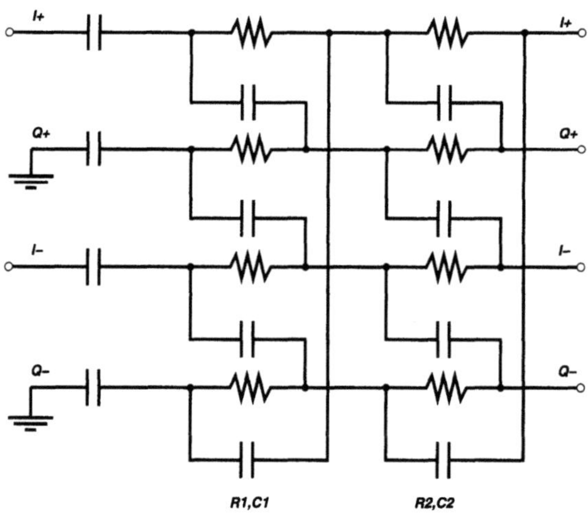

Figure 9.18.   2-stage PPF schematic.

Dual Modulus Prescaler Design

The dual-modulus prescaler (DMP) consists of a divide-by-4/5 synchronous divider and a divide-by-8 asynchronous divider. The architecture of DMP is shown in Figure 9.20. Basic working principal is summarized first, and design and integration issues are discussed next.

When the mode input of the DMP is zero, the output of the second NAND gate in the synchronous divider is always at logic one. Consequently, the

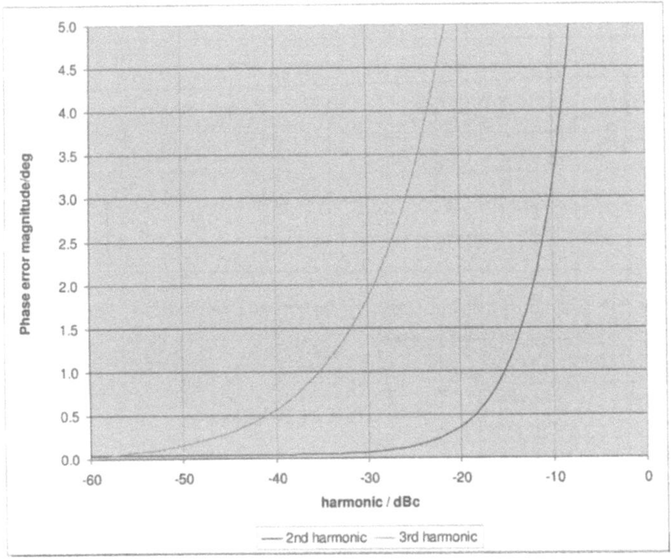

Figure 9.19.   IQ phase error due to 2nd and 3rd harmonic levels at PPF output.

first NAND gate acts as a simple inverter and the first two D-flip-flops form a synchronous divide-by-4 frequency divider. Together with the divide-by-8 asynchronous divider, the total division modulus of the prescaler is 32. When the mode input of the prescaler is at logic one, the input to the second NAND gate becomes one for a brief moment once every output period of the prescaler. During this moment, the output of the second NAND gate is the inverted output of the second D-flipflop, such that the feedback loop is closed over the three D-flip-flops of the synchronous divider. The third flipflop adds an extra delay of exactly one period of fin. This delay is added to the output of the prescaler once every output period, resulting in a division by 33.

Design and integration issues for a typical DMP are listed as;

- All D-flipflops of the synchronous divider operate at the VCO frequency. This increases the clock load to VCO and the power consumption of prescaler.

- Both NAND gates are in the critical path of the synchronous divider. The additional delay causes the maximum operating frequency of the dual-modulus prescaler to be much lower than that of a fixed prescaler (LO1 case). A fixed prescaler operate at the speed of a single high-speed D-flipflop which can be optimized for divide-by-two operation (LO1 case).

- Dual-modulus prescaler can not be driven directly by VCO output for two main reasons; (i) if DMP clock input is connected to VCO output directly,

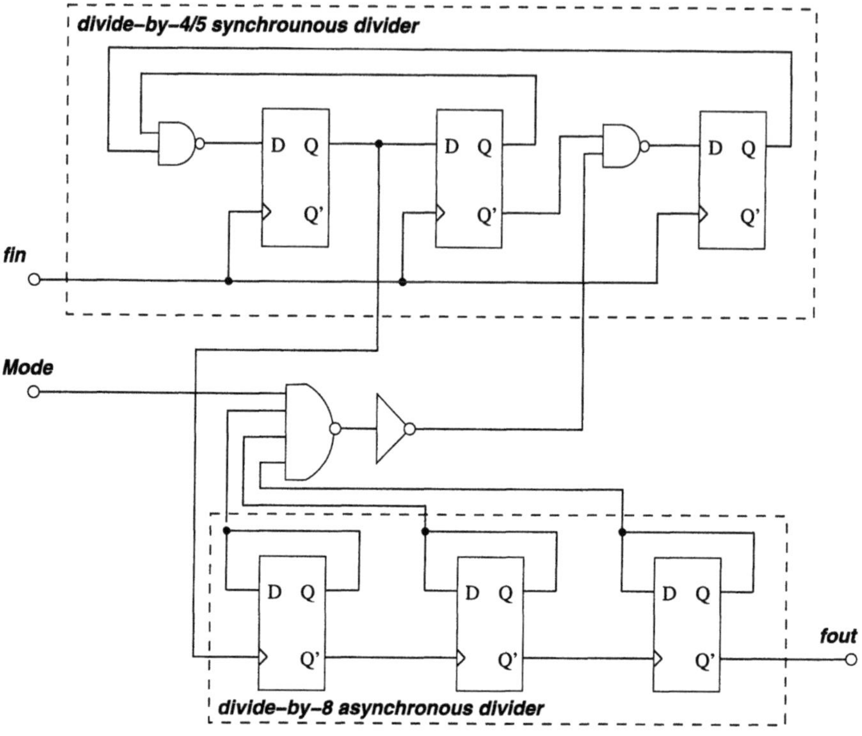

Figure 9.20.   Dual-modulus prescaler architecture.

the DMP will inject switching noise to VCO resonator when DMP changes mode. This noise is also called "kick-back" noise. (ii) It introduces a large capacitive load to VCO tank.

- D-flipflop operating at high-speed is sensitive to input amplitude at the frequency of operation. To guarantee a robust functionality of the DMP, input clock amplitude must be kept greater than the minimum operational level.

First two issues are solved by designing two different D-flipflops; one for the synchronous divider optimized for high-speed and the other for the asynchronous divider to operate at lower frequencies [81]. This way power consumption is reduced. The implemented D-flipflops are based on the true single-phase clocking (TSPC) which are presented in [82, 83]. It can operate up to 2.5GHz, but it requires rail-to-rail single ended clock input. To solve the last two issues, a differential-to-single ended buffer is designed to perform two functions; (i) it converts differential VCO signal as low as 300mVpp to a rail-to-rail signal (ii) it isolates VCO resonator from DMP switching noise and introduces very small load capacitance to VCO resonator, compared to the DMP clock input.

7. Measured Results

The measured integrated phase noise of the frequency synthesizers is less than -34dBc at the PA output which exceeds the 64-QAM performance requirements for 802.11a/g [78]. The untrimmed IQ phase mismatch is 0.3 degree and less than 0.1 degrees after phase trim. The IQ amplitude mismatch is 0.1dB. The total power dissipation by the two LOs is about 30mA from a 1.8V supply voltage. The composite phase noise of the PLLs at the 2.4GHz transmitter output is shown in Figure 9.21 [78].

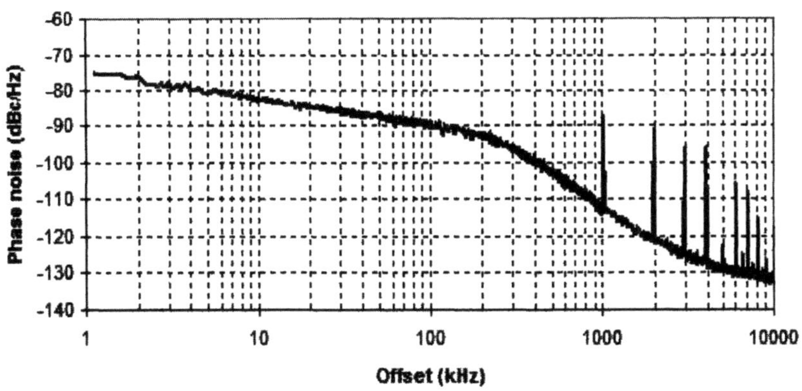

Figure 9.21. Composite phase noise plot of the PLLs at the 2.4GHz transmitter output.

The die photo of LO1 and LO2 frequency synthesizers are shown in Figure 9.22(a) and (b), respectively. The summary of measured results and comparison with published results are listed in Table 9.5.

The 2.4GHz transmitter output spectrum and EVM plots are shown in Figure 9.23 [78]. The 5GHz transmitter output spectrum and EVM plots are shown in Figure 9.24 [78].

8. Summary

A new LO generation architecture is presented for a multi-band and multi-standard (IEEE 802.11a/b/g) fully integrated WLAN radio [78]. The radio architecture has the following advantages; (i) IQ generation is done at lower frequency; and hence good matching values are obtained for IQ phases and

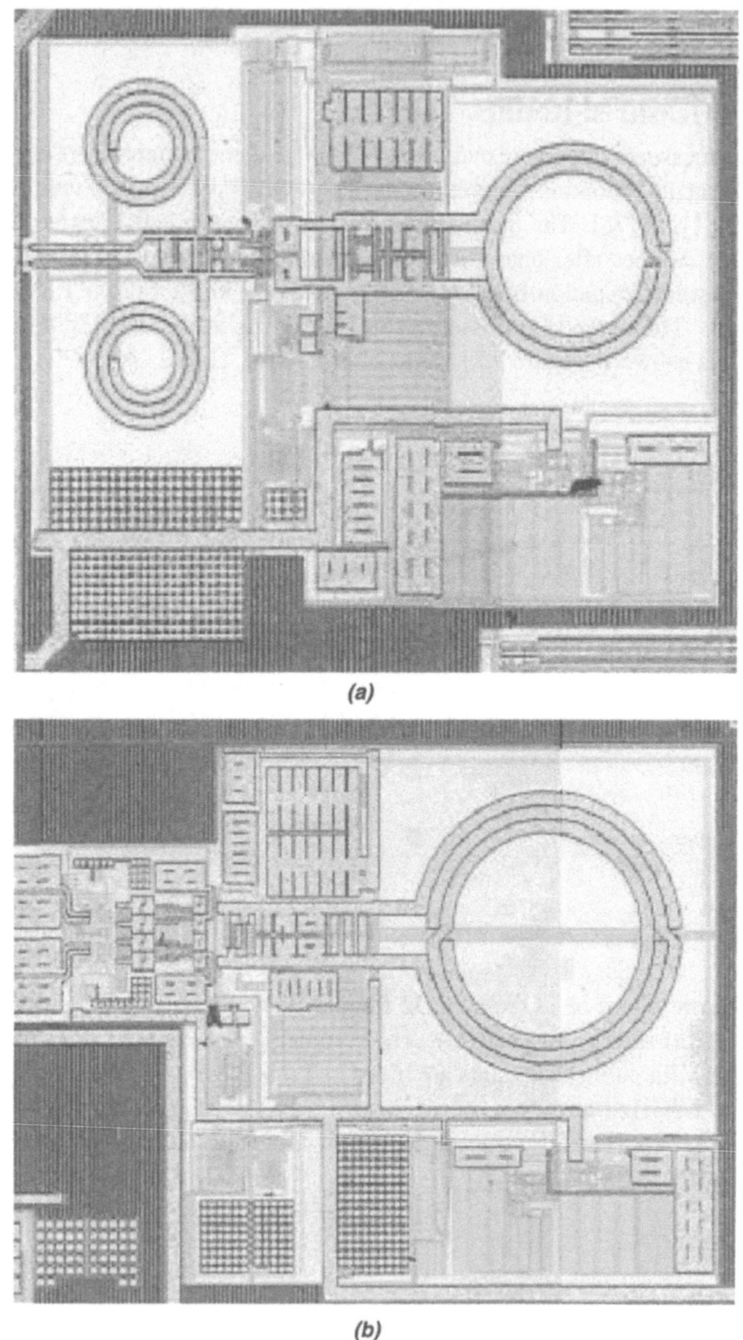

Figure 9.22. Die photo of (a) LO1 (b) LO2.

Parameter	This work	ref [84]	ref [85]
Technology	0.18um CMOS	0.18um CMOS	0.5um SiGe BiCMOS
Supported Stand.	802.11 a/b/g	802.11a	802.11a
Power Consumption (mW)	55	x	145
Integrated Phase noise(dBc)	-34 (-36)	-31.6	-33
IQ phase mismatch (deg)	0.3 (< 0.1)	2	x

Table 9.5. Comparison of this work with published works.

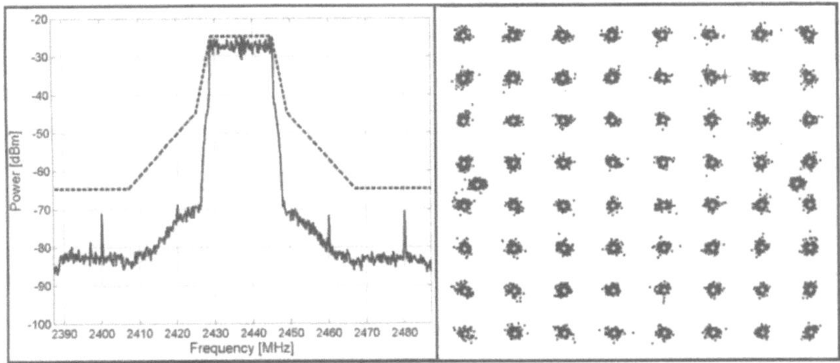

Figure 9.23. 2.4GHz transmitter performance with 64-QAM signal.

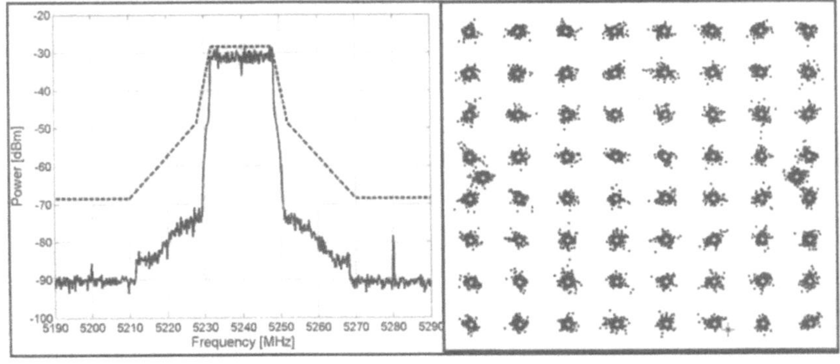

Figure 9.24. 5GHz transmitter performance with 64-QAM signal.

amplitude (ii) channel selection is moved to lower frequencies; and hence the design of dual-mode prescaler is eased and lower power consumption is obtained (iii) RF PLL requires only fixed prescaler, and this greatly simplifies the implementation and reduces power consumption. The two frequency synthesizers, RF and IF, are designed and implemented in $0.18\mu m$ CMOS technology.

Chapter 10

RF CMOS COMPONENT CHARACTERIZATION

The rapid advancement of CMOS technology in the last decade has made possible CMOS implementation of radio frequency (RF) integrated circuits. Commercial fully integrated RF CMOS SoC (System on Chip) implementations of less demanding radio standards such as Bluetooth have started to appear [86]. A major obstacle in CMOS RFIC design is the availability of high quality models for active and passive components at GHz frequencies [87]. Also, the development of a CMOS process is heavily digitally oriented. CMOS foundries first develop the technology for digital design which mainly targets speed and power dissipation. After stabilizing the digital CMOS process, the foundry develops different mixed-signal and analog RF versions of this process by adding masks or modifying the process slightly, e.g. changing the top metal layer to a thick copper layer or adding a special device mask. Recently, some foundries have started to provide special RF CMOS processes derived from digital ones.

A well characterized RF active and passive components model library is needed for a successful RFIC design in order to simulate the designed circuit behavior correctly and to optimize the circuit. The usual approach is to prepare a test chip composed of several active and passive components patterns to characterize the RF devices in the technology. The measurement procedure of test devices and extraction of the models from measurement results are quite important and are prone to many sources of error in the measurement system. Therefore, the measurement of the component is done by microwave wafer probing on a bare silicon die to avoid bond wire, package and fixture effects.

In this chapter, RF CMOS component characterization methodologies are discussed and new techniques for RF CMOS characterization are presented. Measurement and calibration methodologies are of great importance for RF

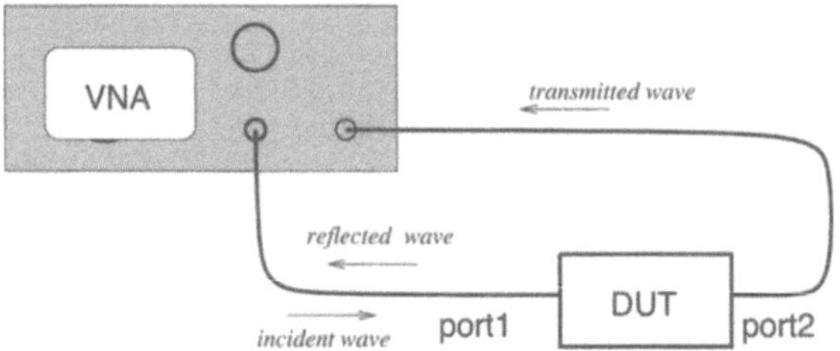

Figure 10.1.   2-port network analyzer measurement.

device characterization. Measurement and calibration methodologies are also discussed.

1.   Calibration and Measurement Techniques

RF components are often characterized by using a vector network analyzer (VNA) which measures vector ratios of reflected and transmitted energy to energy in response to incident upon the device under test (DUT) [88, 89] as shown in Figure 10.1.

A VNA makes a stimulus-response measurement, i.e., while power is applied from one port, the reflected and transmitted power is measured at that port and at the other port, respectively. A VNA measurement determines the properties of the device under test rather than the properties of signals. One must define the boundary where the measurement system ends and the DUT begins in the setup. This boundary is often called 'reference plane'. A reference plane can be established by measuring patterns with known electrical characteristic. This process is called calibration. There are four standard planar patterns: short, open, matched load, and through (SOLT) for on wafer calibration. There are also advanced calibration methods: thru-reflect-line (TRL) and line-reflect-match (LRM) which provide more accurate calibration over SOLT for on wafer probing [88, 90–92]. The calibration patterns can be fabricated on the wafer together with the devices being characterized or on an off-wafer calibration substrate called Impedance Standard Substrate (ISS) [88] which has high precision patterns for calibration. The ISS is usually fabricated on a specific material with a low loss and smooth surface such as ceramic, sapphire. However, the substrate material for calibration patterns is not very critical as long as the calibration standards are well understood and accurately described to the network analyzer correction algorithm [93]. Since calibration is done from probe tips, the reference plane will be the probe tips. The setup will measure the response of whatever is touching the probe tips. The probe tips are usually referenced to the standard level of 50Ω after the calibration procedure.

First, microwave on-wafer probing measurement steps can be summarized as follows [91, 93];

- calibrate the network analyzer up to tips of probe by using either on-wafer or off-wafer calibration standard patterns.

- verify the calibration on the measurement wafer. Verification of the calibration can be done using high-Q inductors or capacitors. Verification will not be correct if it is done on the standard pattern where calibration is done. This is because those patterns are already used for calibration.

- measure the s-parameters of the dummy device and convert them to y-parameters (Appendix A).

- measure the s-parameters of the DUT and convert them to y-parameters.

- subtract the dummy y-parameters from DUT y-parameters, and convert the results back to s-parameters for model library use.

2.   Pad De-embedding

A measurement result from a calibrated probe is the response of the device under test including parasitics associated with probe pads. In order to get the DUT response from measurement, the pad parasitics must be removed. Layout patterns, one including the DUT while the other (dummy) excluding it, are fabricated on the same wafer as shown in Figure 10.2. The correction of measurement results for pad parasitics is often called 'pad de-embedding'. Here, we examine both pad de-embedding and probe pad layout techniques since they are closely related. Proper probe layout rules in addition to technology design rules must be followed [94].

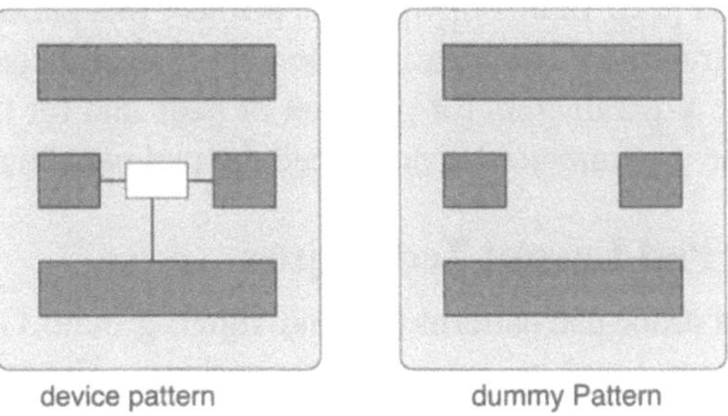

Figure 10.2.   Device and dummy layout.

3.   Pad De-embedding Considerations for RF CMOS

The main interest in RF probing for device characterization purposes is to characterize the intrinsic device for the purpose of modeling its behavior at the GHz frequencies when embedded in an IC design environment. It is obvious that the intrinsic device in an IC design environment will not have probe pads attached to it except when used as a test structure. Therefore, the probe pad parasitic effect must be de-embedded from the measurement since a measurement on wafer with calibrated probe tips has the intrinsic device characteristics plus pad parasitics. There are two approaches for de-embedding pad parasitics. One approach is to place calibration patterns for standards (open, short, load,through) on the same test die with pads, which also mimic the DUT pads, and use these standard patterns for network analyzer calibration [93]. This approach has the advantage of having calibration standards on the same material which is silicon for CMOS as DUT. On the other hand, this approach has a few drawbacks;

1 difficulty in accurately defining the calibration standards with silicon substrates.

2 vertical dimensions vary greatly (up to 30 %) on the surface of CMOS wafers, hence parasitics associated with vertical dimensions cause great variance in the calibration process, i.e., the same standard patterns for calibration fabricated on the same wafer but at different locations may produce different calibration results.

3 since a silicon substrate has finite loss, it will create a coupling parasitics term which will not be accounted for by the network analyzer calibration calculation model.

The second approach is to use off-wafer standards on an Impedance Standard Substrate (ISS) [88]. This method is very practical and is considered the most accurate method [92]. In this method, the primary pad parasitics are in parallel with the pad structure. The idea is to be able to easily convert the measured s-parameters to y-parameters for open set of pads and for the DUT with pads and subtract the y-parameters to de-embed the pad parasitics.

4.   Probe Pad Layout Techniques

Two kinds of probe pad patterns (ground-signal-ground, G-S-G) are shown in Figure 10.3.(a) for 2-port microwave wafer probing . The one in Figure 10.3(a) is highly lossy due to low CMOS substrate resistivity and its one port side is shown in Figure 10.5(a). It can be seen from its cross sectional view in Figure 10.5(b) that the equivalent circuit, Figure 10.5(c), from signal to ground is composed of the parasitic fringe capacitance (C_1) and a lossy $C_2 - R_s - C_3$

path which introduce substrate parasitic effects. This lossy path also causes coupling between port-1 (P1 side) and port-2 (P2 side) as shown in Figure 10.4, i.e., $|s_{21}|$ and $|s_{12}|$ will have non-negligible values. The feed forward parasitic term or coupling term between two ports of characterized devices will have an important impact on the pad de-embedding process and measured device s-parameters. The s_{11} parameter on the Z smith chart is shown in Figure 10.5(d). The one port signal pad equivalent circuit is a lossy RC network as shown in Figure 10.5(c).

The second probe pad pattern uses ground shield underneath the signal pad as shown in Figure 10.3(b). Ground shield is implemented under the signal pad and hence signal leakage is prevented from signal pad to substrate. Ground shield is implemented using the bottom metal layer, M1, under the signal pads as shown in Figure 10.6(a). The M1 layer should be extended in every direction under the signal pad to prevent peripheral fringe capacitance to the substrate.

Ground shield under the signal pad has several benefits in CMOS RF probing. One benefit is that coupling between the probe ports will be minimized compared to the unshielded pad structures since the signal pad is isolated from the substrate. This coupling becomes important especially in a CMOS technology with a low substrate resistivity.

Another benefit is that the process of de-embedding the pad parasitics becomes more simpler because its equivalent circuit is pure capacitive with a high quality factor (Q). The capacitance structure is shown from side view in Figure 10.6(b) and its equivalent circuit in Figure 10.6(c). The s_{11} parameter measurement smith chart plot will, hence, be on the capacitive circle path in the bottom half of the Z smith chart as shown in Figure 10.6(d).

Calibration verification is particularly important when using off-wafer calibration patterns, i.e., ISS with SOLT (short-open-load-thru) calibration to ensure both correct coefficient entry into VNA and successful probe standard measurement. A high Q factor capacitor formed between the signal pad and ground metal can be used in the verification of the calibration.

Another benefit for a shielded probe pad is a better noise characterization of devices because substrate coupling through the signal pad is minimized.

The layout for probe pads can be either G-S-G or just G-S/S-G patterns. The G-S/S-G patterns will occupy less wafer area when considering the fabrication of many devices for modeling purposes. G-S-G has the advantage of a higher isolation between the two signal ports. This is because the EM fields are better terminated due to the extra ground for a G-S-G pattern.

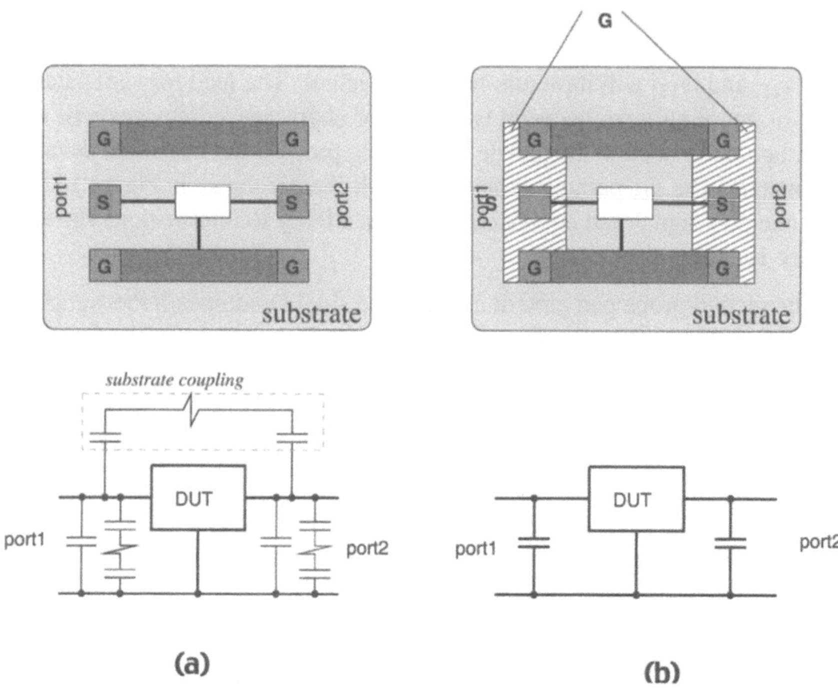

(a)   **(b)**

Figure 10.3.   a) A conventional G-S-G probe pad layout (top view) and a simple equivalent model b) Ground shielded G-S-G probe pad layout and a simple equivalent model.

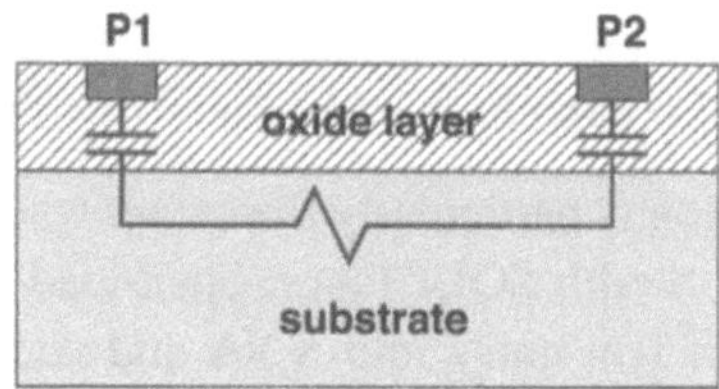

Figure 10.4.   A cross sectional view of the lossy 2-port structure and coupling path between ports through substrate.

5.   Measurement Environment Setups

The measurement setup is described for on wafer probing in this section. Depending on the devices to be measured, different kinds of probes and equipment are required. The equipment needed for measurements are ; a probe station, RF probes , ISS calibration substrate, Vector Network Analyzer, DC power supplies. The device under test (DUT) are bare dies since all measurements are done with on wafer probing. Measurement setups for the test patterns, inductor, varactor, and MOS devices, are listed next.

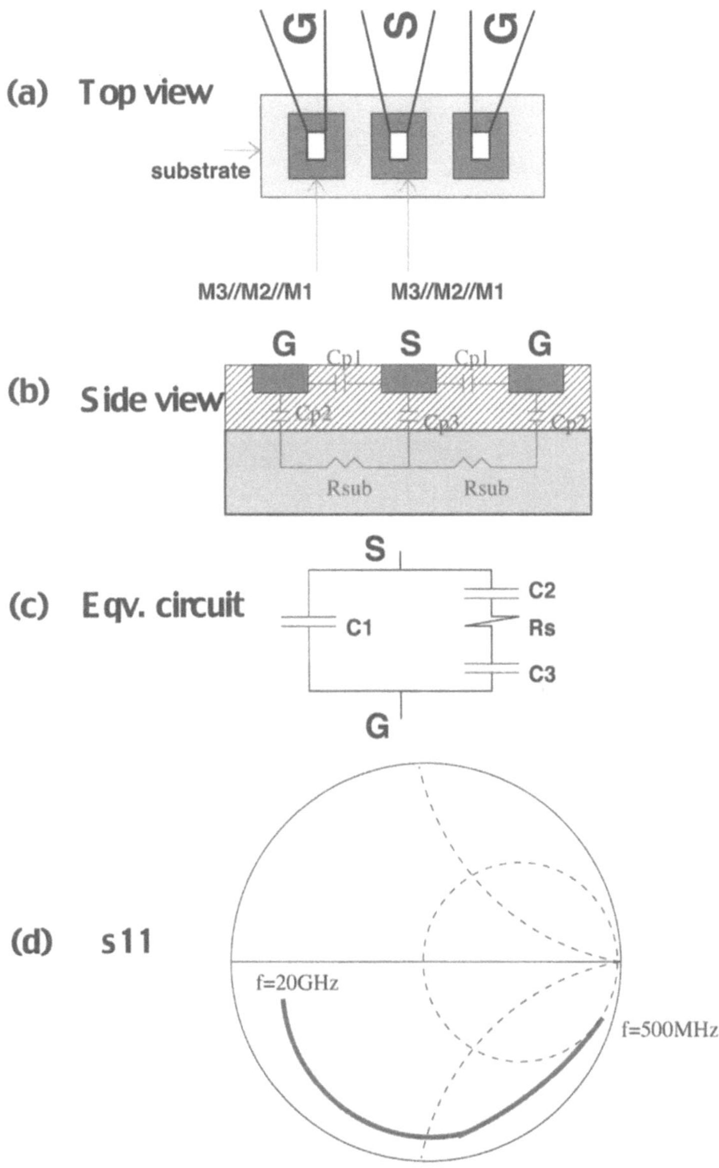

Figure 10.5. a) One port G-S-G top view of the conventional probe pad layout (for 3 metal CMOS process, M1,M2, and M3 are metal layers). b) side view of the layout. c) A simple equivalent circuit for S to G path. d) s_{11} parameter plot on the Z-smith chart.

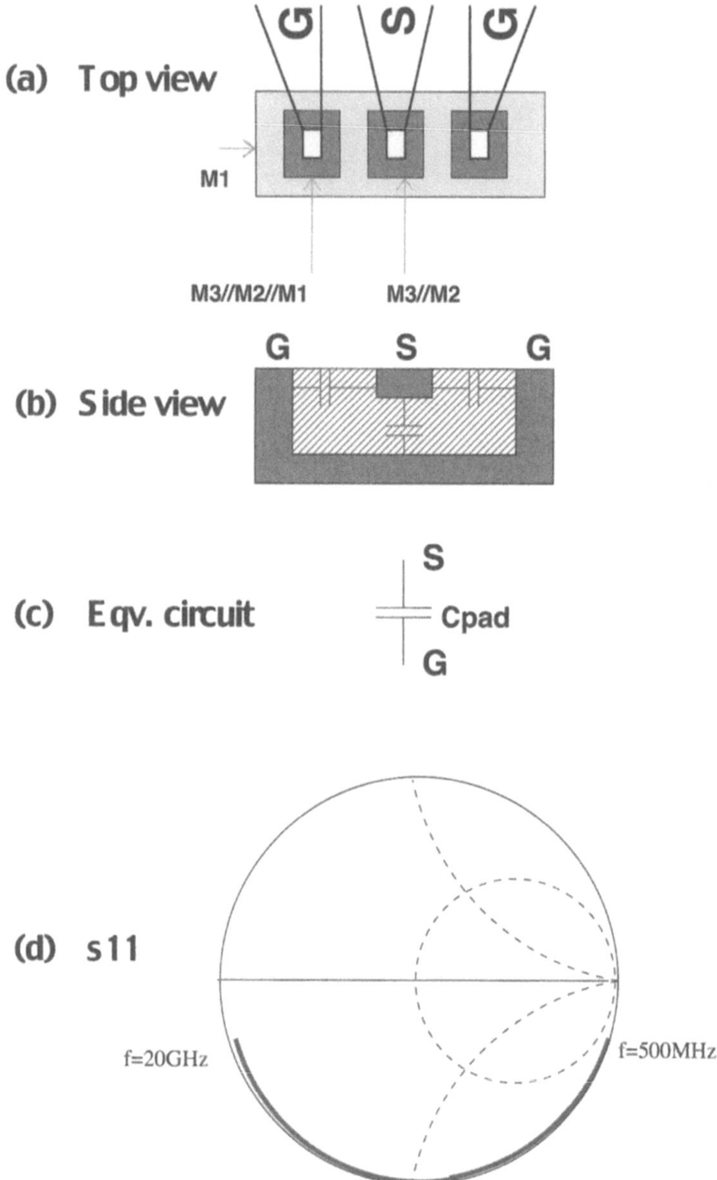

Figure 10.6.   a) One port G-S-G top view of the ground shielded probe pad layout (for 3 metal CMOS process, M1,M2, and M3 are metal layers). b) side view of the layout. c) A simple equivalent circuit for S to G path. d) s_{11} parameter plot on the Z-smith chart.

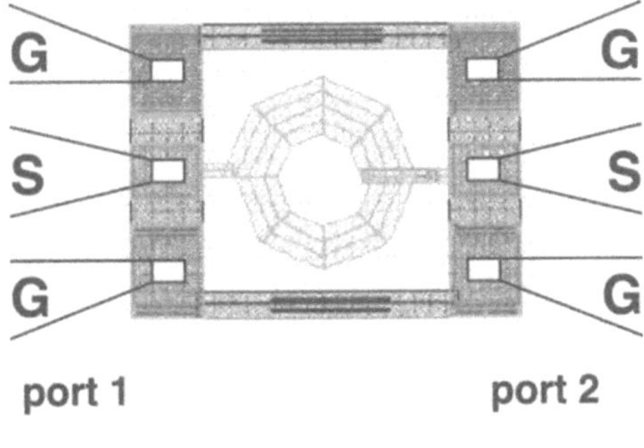

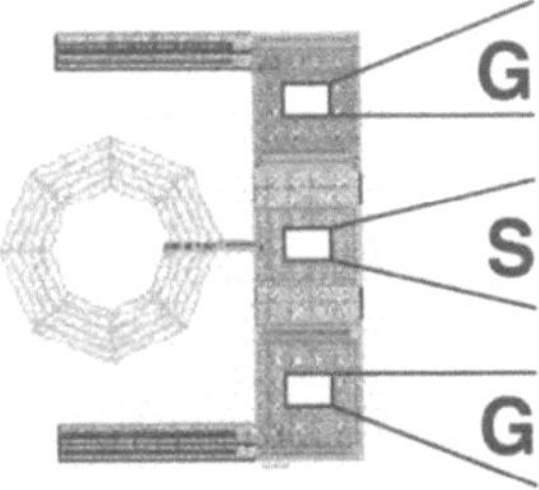

Figure 10.7.   2-port and 1-port inductor measurement setup on die.

RF Inductor Measurement

A network analyzer is needed for 2-port and 1-port s-parameters measurement with RF probe pads. Test Procedure is summarized as following;

1 Calibration - use standard substrate to calibrate probe parasitics up to the tips of probe [93] [94].

2 De-embedding - Measure dummy pads for de-embedding [90].

3 Place the probe as shown in Figure 10.7. Record the s-parameters.

4 Repeat same steps on 4 different dies.

RF Varactor Measurement

A varactors is usually laid out in a differential structure so it requires 2-port measurement. A network analyzer is needed for 2-port s-parameter measurement with RF probe pads. Additional DC voltage sources are also required for biasing purposes. Test Procedure is summarized as following;

1 Calibration - use standard substrate to calibrate probe parasitics up to the tips of probe.

2 De-embedding - Measure dummy pads for de-embedding [90].

3 Place the probe as shown in Figure 10.8.
   - set port-1 bias (Vanod1) - set port-2 bias (Vanod2) - set dc bias needle (Vcontrol)

4 Measure s-parameters for different Vanod and Vcontrol bias points.

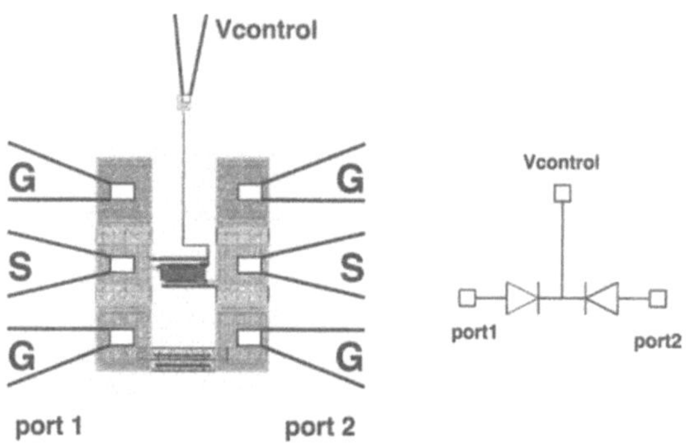

Figure 10.8.   Differential varactor measurement setup.

RF MOS Measurement

A network analyzer is needed for 2-port s-parameters measurement with RF probe pads. Additional DC voltage sources are also required for biasing purposes. Test Procedure is summarized as following;

1 Calibration - use standard substrate to calibrate probe parasitics up to the tips of probe.

2 De-embedding - Measure dummy pads for de-embedding [90].

3 Place the probe as shown in Figure 10.9.
   - set port-1 bias (VG, gate voltage) - set port-2 bias (VD, drain voltage).

4 Measure s-parameters for different VG and VD bias points.

6.   RF CMOS Test Chip

A test chip was designed to characterize CMOS RF components in a 0.5μm CMOS technology. There is no available RF models for this technology. The available small signal model parameters are in BSIM3 format for MOS devices which are verified up to 500MHz [95]. The 0.5μm CMOS process uses a

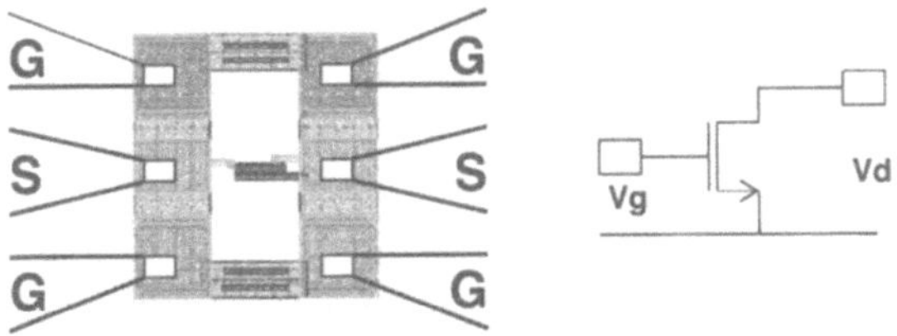

Figure 10.9.   RF MOS device measurement setup.

digitally oriented epi-wafer, i.e., a low doped epi-layer (10 Ω-cm) on highly doped substrate (0.01 Ω-cm). The process has three metal layers and two available poly layers.

The test chip has two types of components, active devices, NMOS and PMOS transistors, and passive ones which include inductors, MOS varactors, diode varactors, and MOS capacitors. The key test patterns on this chip are inductors and varactors for VCO design considerations. Different type inductor structures are laid out. Spiral inductors (square and octagonal) using top single layer (M3), series inductors (using M3 and M2 in series), shunted inductor (M3 and M2 shunted) are laid out to characterize and evaluate different structures. Accumulation mode MOSFET, inversion mode MOSFET and p^+/n junction diode varactor structure are also laid out. The MOS devices, NMOS and PMOS, are also laid out for parameter extraction purposes at low frequency (AC small signal and DC parameters). The RF probe pad is prepared following the guidelines given in References [93] and [94]. A dummy probe pad is also included for probe pad de-embedding purposes. ASITIC [64] has been used for inductor test structure selection. The layout of the test chip is shown in Figure 10.10.

7.   Measurement Results and Technology Evaluation

The measurements were performed on a *Signatone S-1160* probe station. DC-40 GHz probes (Picoprobe Model 40-A GSG-150) were used with the same spacing as the pad structre tested. A vector network analyzer (*R&S ZVC 8GHz*) was used to measure the S-parameters. Picoprobe CS-5 calibration substrates were used for SOLT calibration.

Inductor structures were measured. The active devices in Figure 10.10 were not measurable due to mis-translation of the layout layer forming the n-well during layout submission. Figure 10.11 shows the measured S_{11} of the open dummy pad structure (highlighted as structure 1 in Figure 10.10). The measured Q of the dummy structures is shown in Figure 10.12. It behaves like a low-loss capacitance as expected in Figure 10.6(d). This dummy measurement was used for de-embedding purposes as explained in Section 3.

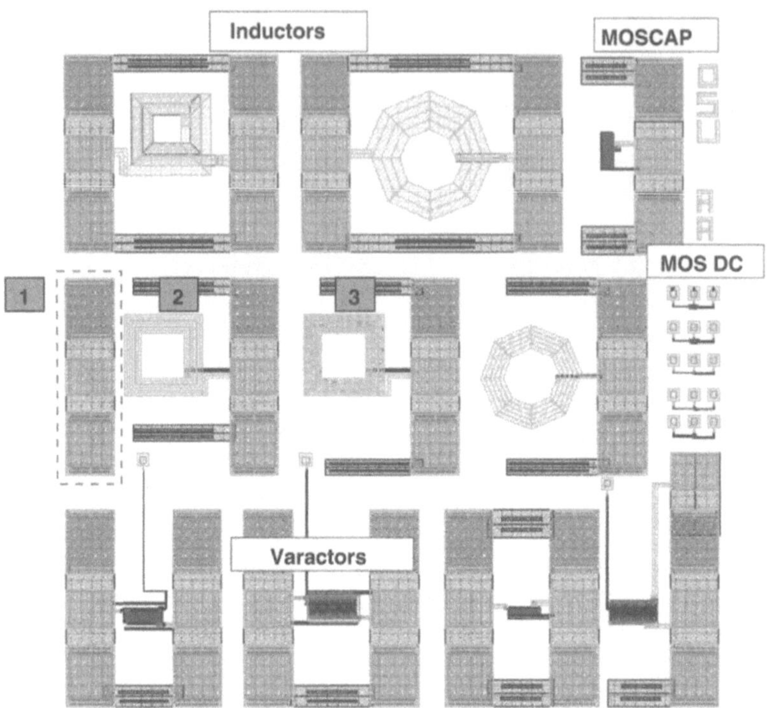

Figure 10.10.   The layout floor plan for the test chip.

Figure 10.13 shows the measured S_{11} of the inductor structure 2 (in Figure 10.10). The extracted Q of the the inductor structure 2 is shown in Figure 10.14. The extracted Q of the the inductor structure 3 is shown in Figure 10.15.

Unfortunately, the evaluated 0.5μm CMOS technology yields low-Q integrated spiral inductors. This is mainly due to the low resistivity substrate (0.01 Ω-cm). Expected inductor Q in this technology is in the range of 3-5. This is also verified with ASITIC simulations. Shunting two or three metal layers in parallel does not improve inductor Q since inductor loss comes mostly from substrate losses, i.e. not from metal resistivity. Historically, early research on spiral inductors for VCO design has focused on elimination of substrate losses. Different techniques were developed for this purposes; selectively removing the underlying substrate with post-fabrication steps [59], using patterned ground shield [60] or use bond-wire instead of integrated spiral inductor [9]. It is concluded here that the evaluated 0.5μm CMOS technology is not suited to design low-phase noise VCOs.

Modern sub-micron CMOS technologies ($0.25\mu m$, $0.18\mu m$, ...) offer a thick top metal layer between $2 - 10\mu m$ on a medium resistivity substrate ($1\Omega \cdot cm < \rho < 20\Omega \cdot cm$). The obtainable Q of integrated inductors in these technologies are in the range of 10-30 at the frequencies 1-5GHz. $0.18\mu m$ CMOS technology with a thick top metal layer of thickness 2μ and substrate resistivity $20\Omega \cdot cm$ is used for WLAN frequency synthesizer and VCO designs.

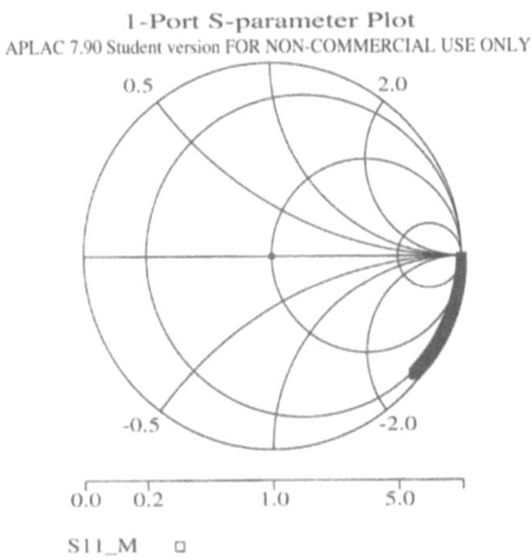

Figure 10.11. S_{11} of the open structure.

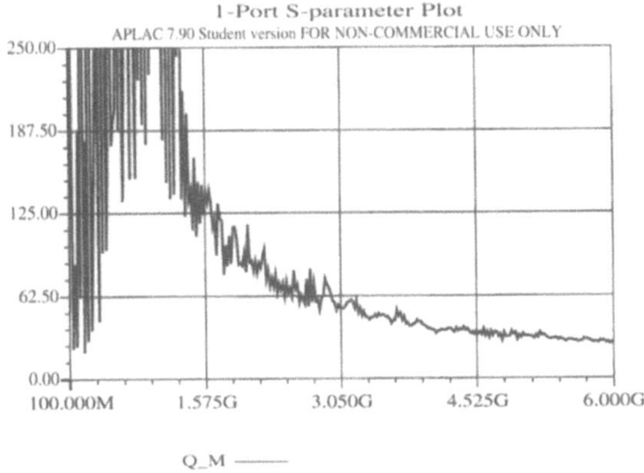

Figure 10.12. Q of the open structure.

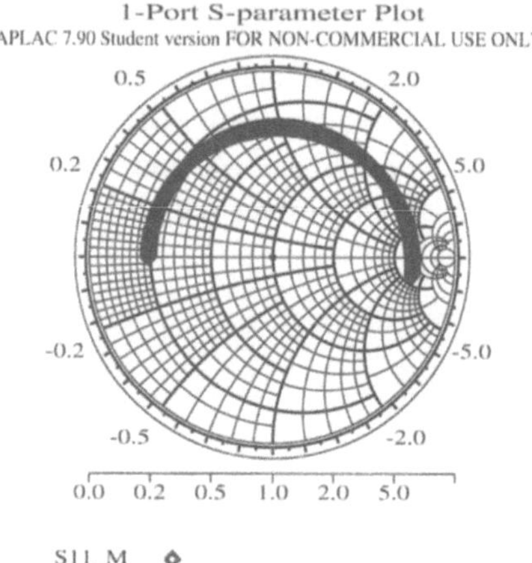

Figure 10.13. S_{11} of the test inductor structure, IND2.

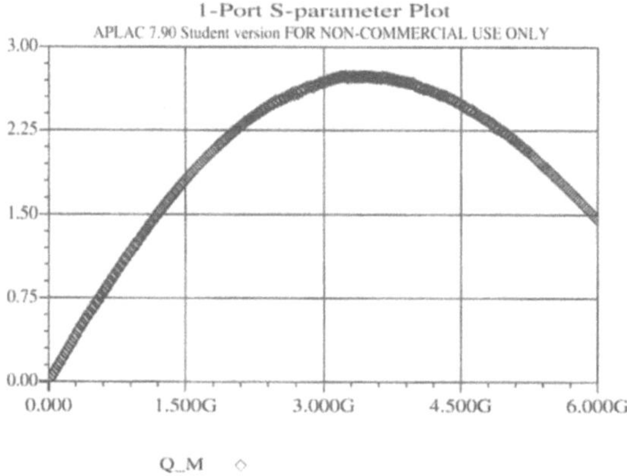

Figure 10.14.   Q of the test inductor structure, IND2.

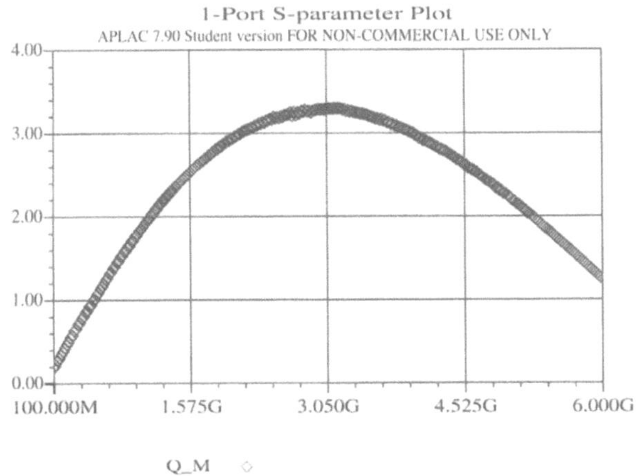

Figure 10.15. Q of the test inductor structure, IND3.

8. Summary

RF CMOS device characterization is described in this chapter. Successful CMOS RFIC design requires a well characterized device library. Foundries provide limited RF models for the devices. Calibration and measurement techniques with microwave probing on wafer are described. Pad de-embedding is also described. A test chip is developed for 0.5um CMOS technology. The 0.5um and 0.18um CMOS technologies are evaluated for RFIC design.

Chapter 11

CONCLUSIONS

In this book, design and integration issues of broadband VCO and PLL frequency synthesizer in CMOS technology are investigated. A fully integrated radio solution requires concurrent design of the transceiver and synthesizer with particular attention paid to frequency planning in order to ease implementation. The full integration of a broadband VCO requires attention at the transceiver architecture and frequency planning, synthesizer architecture, and technology aspects. Summary of conclusions and contributions exploring the above issues are listed as follows:

- An analytical model is developed to investigate the noise properties of closed-loop PLL. The closed-loop PLL noise is an important factor of the total system performance in modern digital communications which use phase modulation, such as QPSK and QAM. The PLL noise model includes noise contributions from the blocks forming a PLL as well as the VCO. A SPICE implementation of the model is also compared with measured results. The models allow optimization of closed-loop PLL noise.

- Broadband VCO design issues are explored. As an individual block, VCO specifications include phase noise, tuning range and power consumption. However, full integration of a VCO in a transceiver also demands manufacturability (robust design against process variation, temperature and supply/bias variations) and tolerance to integrated environment disturbances (substrate coupled noise, cross talk between signal lines, and supply line bounces). Active circuit design for negative resistance generation is explored for broadband operation. The resonator design is also presented.

- An application of broadband PLL frequency synthesizers is designed and implemented for a multi-band/standard (IEEE 802.11a/b/g) WLAN frequency synthesizer in $0.18\mu m$ CMOS. Phase noise trade-offs for PLL noise specifications are explored in this application. A loop filter architecture suited for integrated environments is also developed. VCO interface between prescaler and mixers require particular attention for isolation considerations. A quadrature signal generation architecture is also developed. The quadrature signal is generated by driving PPF with the VCO directly. Dual-modulus prescaler design and integration issues are investigated.

- Design and development of 4GHz VCO in CMOS technology for use in WLAN frequency synthesizer is presented. PMOS and complementary CMOS active circuit topologies are evaluated for integration considerations. PMOS active circuit topology exhibits better performance with accumulation mode varactor.

- An auto calibration circuit for VCO tuning band selection is developed and implemented. The auto calibration circuit eases the tuning range issue for broadband operation.

- A fully integrated dual-mode frequency synthesizer for GSM and WCDMA standards is developed as another application. A dual-band VCO is designed to support the required frequency channels. This example explores hardware sharing in PLL components by using integer-N/fractional-N architectures for dual-band/standard operation.

- RF CMOS characterization issues are investigated. Microwave wafer measurement and pad de-embedding techniques are developed. CMOS technologies are evaluated for RF design.

Other issues which require further study in the design and implementation of fully integrated RF PLL frequency synthesizer in sub-micron CMOS technologies include:

- Amplitude control architecture and circuit techniques for broadband VCOs are needed. VCO output amplitude varies over the tuning range due to change of the resonator Q. Also, inductor metal sheet resistance varies due to temperature and process variations. This results in variation of the Q for the resonator since the loaded resonator Q is determined primarily by the inductor Q.

- Flicker noise of the MOS devices under large signal condition require further research and understanding for better VCO noise prediction.

- Fractional-N architectures for multi-standard operations should be investigated.

- Further investigation of RF CMOS modeling issues at 10GHz and beyond is needed.

Appendix A
S-parameters to Y-parameters Transformation

Measured S-parameters can be transformed to Y-parameters (admittance) by using the following equations.

$$Y_{11} = \frac{(1 - S_{11})(1 + S_{22}) + S_{12}S_{21}}{(1 + S_{11}))1 + S_{22}) - S_{12}S_{21}} \tag{A.1}$$

$$Y_{21} = \frac{-2S_{21}}{(1 + S_{11})(1 + S_{22}) - S_{12}S_{21}} \tag{A.2}$$

$$Y_{12} = \frac{-2S_{12}}{(1 + S_{11})(1 + S_{22}) - S_{12}S_{21}} \tag{A.3}$$

$$Y_{22} = \frac{(1 + S_{11})(1 - S_{22}) + S_{12}S_{21}}{(1 + S_{11})(1 + S_{22}) - S_{12}S_{21}} \tag{A.4}$$

References

[1] Y. Neuvo, "Cellular Phones as Embedded Systems", ISSCC Dig. Tech. Papers,pp.32-37, February 2004.

[2] S.Y.Hui and K.H.Yeung, " Challenges in the Migration to 4G Mobile Systems", *IEEE Communications Magazine*, pp.54-59, December 2003.

[3] M.Shafi, S.Ogose and T. Hattori,*Wireless Communications in the 21st Century*, John Wiley & Sons, 2002.

[4] C. De Ranter and M. Steyaert, *High Data Rate Transmitter Circuits: RF CMOS Design and Techniques for Design Automation*, Kluwer Academic Publishers, 2003.

[5] X.Li and M.Ismail, *Multi-Standard CMOS Wireless Receivers: Analysis and Design*, Kluwer Academic Publishers, 2002.

[6] Adiseno, Mohammed Ismail, and H. Olsson, "A Wide-Band RF Front-end for Multiband Multistandard High-Linearity Low-IF Wireless Receivers", *IEEE J. Solid-State Circuits*, vol. 37, pp. 1162-1168, August 2002.

[7] B. Razavi, "Challenges in the Design of Frequency Synthesizers for Wireless Applications," *Proc. CICC*, pp.395-402, May 1997.

[8] C.S. Vaucher, *Architectures for RF Frequency Synthesizers*, Kluwer Academic Publishers, 2002.

[9] J. Craninckx and M. Steyaert, *Wireless CMOS Frequency Synthesizer Design*, Kluwer Academic Publishers, 1998.

[10] B. Razavi, "A 1.8 GHz CMOS voltage-controlled oscillator," ISSCC Dig. Tech. Papers, pp.388-389, Feb. 1997.

[11] B. Razavi, *RF Microelectronics*. Englewood Cliffs, NJ:Prentice Hall, 1998.

[12] B. Kleveland, C. H. Diaz, D. Vook, L. Madden, T. H. Lee, and S. S. Wong, "Exploiting CMOS Reverse Interconnect Scaling in Multigigahertz Amplifier and Oscillator Design", *IEEE J. Solid-State Circuits*, vol. 36, pp. 1480-1488, October 2001.

[13] G.D. Vendelin, A.M. Pavio, and U.L. Rohde, *Microwave Circuit Design Using Linear and Nonlinear Techniques.* John Wiley & Sons, 1990.

[14] U.L. Rohde, *Microwave and Wireless Synthesizers: Theory and Design.* John Wiley & Sons, 1997.

[15] W.F. Egan, *Frequency Synthesis by Phase Lock*, John Wiley & Sons, 1981.

[16] R.E. Best, *Phase-Locked Loops Theory, Design, and Applications*, 2nd ed., McGraw-Hill Inc, 1993.

[17] B. Razavi,ed. *Monolithic Phase-Locked Loops and Clock Recovery Circuits*,IEEE Press, 1996.

[18] D. Banerjee, *PLL Performance , Simulation and Design*, National Semiconductor, 1998.

[19] D.B. Leeson, "A simple model for oscillator noise spectrum," *Proc. IEEE*, vol. 54, pp. 329 330, Feb. 1966.

[20] K.A. Kouznetsov and R. G. Meyer,"Phase Noise in LC Oscillators," *IEEE J. Solid-State Circuits*, vol. 35, pp. 1244-1248, August 2000.

[21] A. Hajimiri and T.H. Lee, "A General Theory of Phase Noise in Electrical Oscillators," *IEEE J. Solid-State Circuits*, vol.33, pp.179-194, February 1998.

[22] A.Demir, A. Mehrotra, and J. Roychowdhury, "Phase Noise in Oscillators: A Unifying Theory and Numerical Methods for Characterization", *IEEE Trans. on Circuits and Systems-I:Fundamental Theory and Applications,* vol.47 pages:655-674, May 2000.

[23] www.mentor.com/ams/eldorf.html

[24] www.cadence.com/datasheets/spectrerf.html

[25] eesof.tm.agilent.com

[26] T. A. D. Riley, M. A. Copeland, and T. A. Kwasniewski, "Delta-sigma modulation in fractional-N frequency synthesis," *IEEE J. Solid-State Circuits*, vol. 28, pp. 553-559, May 1993.

[27] A. Aktas, F.Johnson, R.Ahola, and M.Ismail, "A 4 GHz 0.18μm CMOS PLL Frequency synthesizer with wide-band VCO for multi-standard wireless applications," *Proc. of NORCHIP 2003*, Nov. 2003.

[28] S.Li, I.Kipnis and M.Ismail, "A 10-GHz CMOS Quadrature LC-VCO for Multirate Optical Applications", *IEEE J. Solid-State Circuits*, vol.38, pp.1626-1634, October 2003.

[29] Y.Wu and V.Aparin, " A Monolithic Low Phase Noise 1.7GHz CMOS VCO for Zero-IF Cellular CDMA Receivers", ISSCC Dig.Tech. Papers, pp.396-397, Feb. 2004.

[30] A.A. Abidi, "Direct-conversion radio transceivers for digital communication," *IEEE J. Solid-State Circuits*, vol.30, pp.1399-1410, December 1995.

[31] J. Strange and S. Atkinson, " A direct conversion transceiver for multi-band GSM application," *In 2000 RFIC Symp. Dig. Papers*, pp.25-28, 2000.

[32] A. Zolfaghari and B. Razavi, "A Low-Power 2.4-GHz Transmitter-Receiver CMOS IC," *IEEE J. Solid-State Circuits*, vol.38, pp.176-183, February 2003.

[33] Y. Tang, A. Aktas, M. Ismail, S. Bibyk, " A fully integrated dual-mode frequency synthesizer for GSM and Wideband CDMA in $0.5\mu m$ CMOS", *Proc. of the 44th IEEE 2001 Midwest Symposium*, Vol. 2, pp.866-869, 2001.

[34] A. Kral, F. Behbahani, and A.A. Abidi, "RF-CMOS oscillators with switched tuning", *Proceedings of Custom Integrated Circuits Conference*, pp. 555-558, Santa Clara, CA May, 1998.

[35] J.F. Parker and D. Ray, "A 1.6-GHz CMOS PLL with On-Chip Loop Filter," *IEEE J. Solid-State Circuits*, vol. 30, pp. 1457-1462, Mar. 1998.

[36] C.Lam and B. Razavi, "A 2.6-GHz/5.2-GHz Frequency Synthesizer in 0.4-um CMOS Technology," *IEEE J. Solid-State Circuits*, vol. 35, pp. 788-794, May 2000.

[37] J.W.M. Rogers , J.A. Macedo and C. Plett, "The effect of varactor nonlinearity on the phase noise of completely integrated VCOs,"*IEEE J. Solid-State Circuits*, vol. 35, pp. 1360-1367, September 2000.

[38] C. Samori, A. L. Lacaita, A. Zanchi, S. Levantino, and G. Cali,"Phase noise degradation at high oscillation amplitudes in LC-tuned VCO's," *IEEE J. Solid-State Circuits*, vol. 35, pp. 96-99, Jan. 2000.

[39] C. Samori, A. L. Lacaita, F. Villa, and F. Zappa, "Spectrum folding and phase noise in LC tuned oscillators," *IEEE Trans. Circuits Syst. II*, vol.45, pp. 781-790, July 1998.

[40] C. Samori, A. L. Lacaita, A. Zanchi, S. Levantino, and F. Torrisi,"Impact of indirect stability on phase noise performance of fully integrated LC tuned VCOs,"*in Proc. Eur. Solid-State Circuits Conf.*, Duisburg, Germany, pp. 202-205, Sept. 1999.

[41] D. F. Peterson, "Varactor properties for wide-band linear-tuning microwave VCOs," *IEEE Trans. Microwave Theory Tech.*, vol. MTT-28, pp. 110-119, Feb. 1980.

[42] E.Hegazi and A.A.Abidi,"Varactor Characteristics, Oscillator Tuning Curves, and AM-FM Conversion" '*IEEE J. Solid-State Circuits*, vol. 38, pp. 1003-1009, June 2003.

[43] S. Levantino, C. Samori, A. Bonfanti, S. L. J. Gierkink, A. L. Lacaita, and V. Boccuzzi, "Frequency dependence on bias current in 5-GHz CMOS VCOs: Impact on tuning range and flicker noise up-conversion," *IEEE J. Solid-State Circuits*, vol. 37, pp. 1003-1011, Aug. 2002.

[44] T. Soorapanth, C.P. Yue, D.K. Shaeffer, T.H. Lee and S.S. Wong, "Analysis and Optimization of Accumulation-Mode Varactor for RF ICs", Symposium on VLSI Circuits, pp. 32-33, June 1998.

[45] A.S. Porret, et al., "Design of High-Q Varactors for low-power Wireless Applications", *IEEE J. Solid-State Circuits*, Vol.35 , pp.337-345, March 2000.

[46] R. Castello, P. Erratico, S. Manzini and F. Svelto, "A $\pm$ 30% Tuning Range Varactor Compatible with Future Scaled Technologies" , *1998 Symposium on VLSI Circuits Digest of Technical Papers*, pp. 34-35, June 1998.

[47] APN1007, "Switchable Dual-Band 170/420 MHz VCO for Handset Cellular Applications," http://www.alphaind.com/

[48] E. Pedersen,"RF CMOS Varactors for 2 GHz Applications", *Analog Integrated Circuits and Signal Processing*, Vol.26, pp.27-36, January 2001.

[49] P. Andreani and S. Mattisson, "On the use of MOS varactors in RF VCOs,"*IEEE J. Solid-State Circuits*, vol. 35, pp. 905-910, June 2000.

[50] P. Andreani and S. Mattisson, "A 2.4-GHz CMOS monolithic VCO based on an MOS varactor,"*in Proc. ISCAS'99, vol. II*, pp. 557-560. May/June 1999.

[51] S.Kim, K.K. O,"Demonstration of a Switched Resonator Concept in a dual-band monolithic CMOS LC-tuned VCO," *Proc. of the Custom Integrated Circuits Conference*, pp.205-208. May 2001.

[52] M.Tibeout, "A CMOS Fully Integrated 1 GHz and 2 GHz Dual Band VCO with a Voltage Controlled Inductor,"*Proc. ESSCIRC 2002*, pp.799-802. 2002.

[53] A. Hajimiri and T. H. Lee, "Design issues in CMOS differential LC oscillators," *IEEE J. Solid-State Circuits*, vol. 34, pp. 717 724, May 1999.

[54] J.W.M. Rogers, D.Rahn, and C.Plett, "A Study of Digital and Analog Automatic-Amplitude Control Circuitry for Voltage-Controlled Oscillators", *IEEE J. Solid-State Circuits*, vol. 38, pp. 352-356, January 2003.

[55] M. Margarit, J. Tham, R. Meyer, and M. Deen, "A low-noise low-power VCO with automatic amplitude control for wireless applications,"*IEEE J. Solid-State Circuits*, vol. 34, pp. 761-771, June 1999.

[56] J. Long and M. Copeland, "The modeling, characterization, and design of monolithic inductors for silicon RF IC's,"*IEEE J. Solid-State Circuits*, vol. 32, pp. 357-369, Mar. 1997.

[57] J. N. Burghartz, D. C. Edelstein, K. A. Jenkins, and Y. H. Kwark,"Spiral inductors and transmission lines in silicon technology using copper damascene interconnects and low-loss substrates,"*IEEE Trans. Microwave Theory Tech.*, vol. 45, pp. 1961-1968, Oct. 1997.

[58] A. C. Reyes, S. M. El-Ghazaly, S. J. Dorn, M. Dydyk, and D. K. Schroder, "Coplanar waveguides and microwave inductors on silicon substrates," *IEEE Trans. Microwave Theory Tech.*, vol. 43, pp. 2016-2022, Sept. 1995.

[59] J. Y.-C. Chang, A. A. Abidi, and M. Gaitan, "Large suspended inductors on silicon and their use in a 2-μm CMOS RF amplifier," *IEEE Electron Device Lett.*, vol. 14, pp. 246-248, May 1993.

[60] C. P. Yue and S. S. Wong, "On-chip spiral inductors with patterned ground shields for Si-based RF IC's,"*IEEE J. Solid-State Circuits*, vol. 33, pp. 743-752, May 1998.

[61] Momentum/Planar EM Simulator, http://eesof.tm.agilent.com/products/e8921a-a.html

[62] FEMLAB,http://www.femlab.com/electro/

[63] SONNET, http://www.sonnetusa.com/

[64] ASITIC, http://formosa.eecs.berkeley.edu/ñiknejad/asitic.html

[65] A. M. Niknejad, R. G. Meyer, "Analysis, design, and optimization of spiral inductors and transformers for Si RF ICs", *IEEE J. Solid-State Circuits*, vol.33, pp.1470-1481, Oct. 1998.

[66] S.S. Mohan, M.Hershenson, S.P.Boyd, and T.H.Lee,"Simple accurate expressions for planar inductances," *IEEE J. Solid-State Circuits*, vol. 34, pp. 1419-1424, Oct. 1999.

[67] R. B. Merrill, T. W. Lee, H. You, R. Rasmussen, L. A. Moberly, " Optimization of High Q Integrated Inductors for Multi-Level Metal CMOS", *Proceedings of the International Electronic Device Meeting*, pp. 983-6, 1995.

[68] M. Danesh, J. R. Long, R. A. Hadaway, and D. L. Harame,"A Q-factor enhancement technique for MMIC Inductors," *IEEE Radio Frequency Integrated Circuits Symp.*, p.217-220, June 1998.

[69] P. Andreani, "A comparison between two 1.8 GHz CMOS VCO's tuned by different varactor," Proc. 24th European Solid-State Circuits Conf., Sep. 1998, pp. 380-383.

[70] W.B. Wilson, U.K. Moon, K.R.Lakshmikumar, and L. Dai, "A CMOS Self-calibrating Frequency Synthesizer", *IEEE J. Solid-State Circuits* Vol 35, pp.1437-1444, October 2000.

[71] Y.Shibahara and M. Kokubo, "A 0.7-V 200-MHZ Self-Calibration PLL",*IEICE Trans. Elec.*, Vol.E85-C, No.8, pp.1577-1580, August 2002.

[72] J. Fenk, "Highly integrated RF-IC's for GSM and DECT systems-A status review," *IEEE Microwave Theory and Techniques*, vol. 45, pp. 2531-2539, Dec 1997.

[73] Murata VCO data sheets, "http://www.murata.com".

[74] IEEE Wireless LAN standards, http://standards.ieee.org/getieee802/802.11.html.

[75] http://www.abiresearch.com/reports/WLAN.html

[76] A.Aktas, K.Rama Rao, J.Wilson and M.Ismail, "A single chip radio transceiver for 802.11 a/b/g WLAN in 0.18um CMOS", Proc. of 10th IEEE International Conference on Electronic Circuits and Systems, ICECS 2003, Dec. 2003.

[77] A.Aktas, K.Rama Rao, J.Wilson, "A tri-band radio architecture for a WLAN transceiver", Sweden PRV. Patent No. 0102554-3. Mar. 2003.

[78] R. Ahola, A. Aktas, J. Wilson, K.Rama Rao, et al., "A Single Chip CMOS Transceiver for 802.11a/b/g WLANs", ISSCC Dig. Tech. Papers,pp.92-93,Feb. 2004.

[79] B. Come, R. Ness, S. Donnay, L. Van der Perre, W. Eberle, P. Wambacq, M. Engels and I. Bolsens, "Impact of front-end non-idealities on Bit Error Rate performances of WLAN-OFDM transceivers," *Microwave Journal*, February 2001.

[80] B. Razavi, K.F. Lee, and R.H. Yan, " Design of High-Speed, Low-Power Frequency Dividers and Phase-Locked Loops in Deep Submicron CMOS," *IEEE J. Solid-State Circuits* Vol 30, pp.101-109, February 1995.

[81] Y.Tang, A.Aktas, M. Ismail and S. Bibyk, "A High-Speed Low-Power Divide-by-15/16 Dual Modulus Prescaler in 0.6μmm CMOS", *Analog Integrated Circuits and Signal Processing* , vol.28, pp. 195-200, August 2001.

[82] J. Yuan and C. Svensson, "High-speed CMOS Circuit Technique," *IEEE J. of Solid-State Circuits*, Vol.24, no.1, pp. 62-70, January 1989.

[83] J. Yuan and C. Svensson, "New single-clock CMOS latches and flipflops with improved speed and power savings," *IEEE J. of Solid-State Circuits*, vol. 32, no. 1, pp. 62-67, January 1997.

[84] P. Zhang, T. Nguyen, C. Lam, D. Gambetta, C. Soorapanth, C. Baohong, S. Hart, I. Sever, T. Bourdi, A. Tham, Razavi, B., "A direct conversion CMOS transceiver for IEEE 802.11A WLANs," *ISSCC Digets of Technical Papers*, pp.1-10, Feb. 2003.

[85] M.A. Margarit,D.Shih, P.J.Sullivan, and F. Ortega, "A 5-GHz BiCMOS RFIC Front-End for IEEE 802.11a/HiperLANWireless LAN," *IEEE J. Solid-State Circuits* Vol 38,pp.1284-1287, July 2003.

[86] F. Op't Eynde, et al., "A Fully-Integrated Single-Chip SOC for Bluetooth," *ISSCC Digets of Technical Papers*, pp.196-197, Feb. 2001.

[87] B. Razavi, "CMOS Technology Characterization for Analog and RF Design," *IEEE J. Solid-State Circuits*, vol.34 pp.268-276, March 1999.

[88] "On-Wafer Vector Network Analyzer Calibration and Measurements," Application Note, Cascade Microtech, Inc.

[89] A. Aktas and M. Ismail, "Pad de-embedding in RF CMOS", *IEEE Circuits and Devices Magazine*, vol.17, pp. 8-11, May 2001.

[90] H. Cho and D.E. Burk, " A Three-Step Method for the De-Embedding of High- Frequency S-parameter Measurements", *IEEE Trans. on Electronic Devices*, Vol. 38, No. 6, pp. 1371-1375, June 1991.

[91] P. J. van Wijnen, et al, "A New Straightforward Calibration and Correction Procedure for On wafer High Frequency S-parameter Measurements (45 MHz-18 GHz)," *Bipolar Circuits and Technology Meeting*, pp. 70-73, 1987.

[92] Arthur Fraser, Reed Gleason, E. W. Strid, "GHz On-Silicon Wafer Probing Calibration Methods" *Bipolar Circuits and Technology Meeting*, pp. 154-157, 1988.

[93] "Introduction to bipolar device GHz measurement techniques," Application Note, Cascade Microtech, Inc.

[94] "Layout Rules for GHz-Probing," Application Note, Cascade Microtech, Inc.

[95] MOSIS Spice Model Parameters, http://www.mosis.org/Faqs/faq-spice.html.

Index

AM-FM conversion, 41
AM noise, 19, 41
Auto calibration, 115
Broadband VCO, 47
  with sub-bands, 47
Calibration, 146
Capacitance switching, 48
Center-tapped inductor, 66
Differential inductor, 66
Differential line, 87
Dual-modulus prescaler, 94
Frequency planning, 37
GSM, 133
Heterodyne, 5
IEEE 802.11a, 73
IEEE 802.11b, 73
IEEE 802.11g, 73
Inductance switching, 56
Integer-N, 138
Jitter, 8
Loop filter, 78
  off-chip, 78
  on-chip, 78
On-wafer probing, 146
Oscillator, 6
Oscillator noise model, 21
Pad de-embedding, 147
Phase noise
  definition, 17
  integrated, 16
  oscillator, 7, 20
  PLL, 23
  specification, 14
  SSB, 26
PLL, 8
  charge Pump, 8
  close-in noise, 26
  dynamic behavior, 12

frequency synthesis, 8
linear model, 9
lock time, 12
Loop filter, 11
noise floor, 26
noise simulation, 28
noise transfer functions, 24
oepn-loop bandwidth, 12
phase margin, 12
PM noise, 19, 41
Polyphase filter, 93
Prescaler, 83
Probe pad, 148
Program counter, 141
Radio architecture, 37
RF switch, 48
Spectral purity, 40
Spiral inductor, 63
Spurious tone, 18
Swallow counter, 141
Time-division duplexing, 124
Tuning, 7
  nonlinearity, 7
Varactor, 67
VCO
  active circuit, 57
  bias, 58
  bias noise filtering, 61
  buffer, 86
  design trade-offs, 58
  programable bias, 60
  pulling, 124
  resonator, 63
  start-up, 60
  topologies, 58
Voltage-controlled Oscillator, 6
WCDMA, 133